Principles of Multimessenger Astronomy

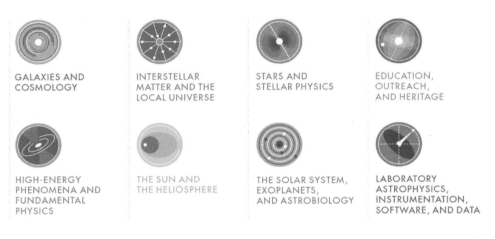

AAS Editor in Chief

Ethan Vishniac, Johns Hopkins University, Maryland, USA

About the program:

AAS-IOP Astronomy ebooks is the official book program of the American Astronomical Society (AAS), and aims to share in depth the most fascinating areas of astronomy, astrophysics, solar physics and planetary science. The program includes publications in the following topics:

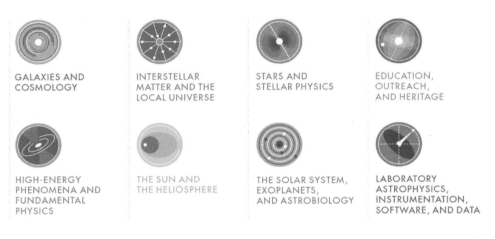

GALAXIES AND COSMOLOGY

INTERSTELLAR MATTER AND THE LOCAL UNIVERSE

STARS AND STELLAR PHYSICS

EDUCATION, OUTREACH, AND HERITAGE

HIGH-ENERGY PHENOMENA AND FUNDAMENTAL PHYSICS

THE SUN AND THE HELIOSPHERE

THE SOLAR SYSTEM, EXOPLANETS, AND ASTROBIOLOGY

LABORATORY ASTROPHYSICS, INSTRUMENTATION, SOFTWARE, AND DATA

Books in the program range in level from short introductory texts on fast-moving areas, graduate and upper-level undergraduate textbooks, research monographs and practical handbooks.

For a complete list of published and forthcoming titles, please visit iopscience.org/books/aas.

About the American Astronomical Society

The American Astronomical Society (aas.org), established 1899, is the major organization of professional astronomers in North America. The membership (~7,000) also includes physicists, mathematicians, geologists, engineers and others whose research interests lie within the broad spectrum of subjects now comprising the contemporary astronomical sciences. The mission of the Society is to enhance and share humanity's scientific understanding of the universe.

Principles of Multimessenger Astronomy

Miroslav D Filipović and Nicholas F H Tothill
School of Science, Western Sydney University, Penrith, NSW 2751, Australia

IOP Publishing, Bristol, UK

ISBN   978-0-7503-2340-6 (ebook)
ISBN   978-0-7503-2338-3 (print)
ISBN   978-0-7503-2341-3 (myPrint)
ISBN   978-0-7503-2339-0 (mobi)

DOI   10.1088/2514-3433/ac087e

Version: 20210901

AAS–IOP Astronomy
ISSN 2514-3433 (online)
ISSN 2515-141X (print)

British Library Cataloguing-in-Publication Data: A catalogue record for this book is available from the British Library.

Published by IOP Publishing, wholly owned by The Institute of Physics, London

IOP Publishing, Temple Circus, Temple Way, Bristol, BS1 6HG, UK

US Office: IOP Publishing, Inc., 190 North Independence Mall West, Suite 601, Philadelphia, PA 19106, USA

Cover image: Artist's impression of a blazar, a type of active galaxy with one of its jets pointing toward Earth. Image credit: IceCube/NASA.

I dedicate this book to my dearest Filipović family: Tatjana, Sofija, Jana, Miloš Koviljka, and Dragoslav. I also dedicate this book to my late grandfather Milan Pere Filipović (1878–1959) who courageously fought in the Great War for the freedom and pride of allies. Finally, I dedicate this book to a fantastic childhood memory of my late aunt Milka Polovina (1918–1996) who is my inspirational hero from World War II.

In Serbian: *Ovu knjigu posvećujem mojoj divnoj Filipović porodici: Tatjani, Sofiji, Jani, Milošu, Koviljki, i Dragoslavu. Takođe, ponosno je posvećujem i mom pradedi Milan-u Pere Filipović-u (1878–1959) za njegovo herojstvo u Velikom ratu. Na kraju, povećujem ovu knjigu i mojoj tetki—Milka Polovina (1918–1996)—mom inspirativnom heroju iz II svetskog rata.*

To Megan, Aidan, and Callum, who put up with the writing of this book; and to the many colleagues, teachers, and students who have given up their time to explain its contents to me over the years.

Contents

Appendices

Preface

Observational Astrophysics—the profession and topic of the authors—tends to suffer from a gap between the knowledge and competencies that can be expected from a new graduate and those that are required to work at the forefront of research. This gap is traditionally bridged by a combination of formal graduate education and oral tradition. But the gap gets broader and longer as astrophysics assimilates more techniques and research progresses. New graduates come into the field from different backgrounds, such as computer science. Multimessenger astronomy demands that students and academics, raised with one set of tools, understand and collaborate with those who were raised with a very different set.

This book and its companion volume—*Multimessenger Astronomy in Practice*—are our response to these challenges. For the second book, we have invited experts in their fields to write an orientation for the newcomer to their topic, whether a new graduate or an established academic. This volume serves as an (extended) introduction to the second book and as an exploration of foundation topics that are useful across messenger and wave band boundaries. We seek to cover the basic principles of physics underlying observational astrophysics, together with scientific history, the operation and use of modern astronomical instrumentation, and the analysis of data obtained from the instrumentation. In particular, we seek to introduce the use of data from multiple messengers (photons, cosmic rays, gravitational waves, and neutrinos) in order to build up the best possible picture of astrophysical objects and processes. Our goal is a useful general reference for all astronomers interested in multimessenger observational astronomy and astrophysics, as well as a teaching resource.

Much of this material was originally created for teaching both undergraduate and postgraduate level units at Western Sydney University (WSU) and elsewhere. At the same time, our postgraduate students (more than 50 over the past two decades) have found the multimessenger approach essential to their education and research.

We begin by presenting a short overview of the history of observational astronomy and astrophysics in the introductory Chapter 1. The following three chapters (2–4) then present a survey of basic concepts of the measurement of astrophysical messengers. We have generally avoided derivations from first principles, and we expect that many readers will already be familiar with much of the content. But we also anticipate that most readers will find something new. In Chapter 5, we detour from basic physics to survey the structure of the Earth's atmosphere, and its effects on astronomical measurement. Chapter 6 introduces emission mechanisms of cosmic messengers, and Chapters 7 and 8 expand our view into particle astrophysics and gravitational waves respectively. Finally, in Chapter 9 we introduce basic principles of astronomical data sets.

We have tried to balance a reasonably comprehensive treatment with a reasonable length; so we have had to omit many topics. Our work concentrates on observational astronomy and astrophysics; the vital work of theoretical astrophysics and of instrumentation specialists is only covered to the extent that is required for an observer. Observational cosmology based on the measurement of the cosmic microwave background is hardly mentioned. No doubt the reader will disagree with at least some of our inclusions and exclusions; we hope that our work will still be found useful.

Acknowledgements

We acknowledge invaluable help and support from a number of our colleagues and students. Namely, we thank:

Dr Jeffrey L. Payne, Dr Howard Leverenz, Dr Clancy James, Velibor Velović, Devika Shobhana, Albany Asher, Adeel Ahmad, Farhan Sher, Rami Alsaberi, Miranda Yew, Dr William C. Millar, Dr Luke Barnes, Dr Paul D. Lasky, Prof. Gawin Rowell, Aleksandar Zorkić, Prof. Bärbel Koribalski, and Dr Ivan Bojičić.

Author Biographies

Miroslav D Filipović

Professor Miroslav D. Filipović is a scientist, philosopher, and philanthropist with over 30 years of experience in astronomy. Astronomy, science, philosophy, and computing are his profession, hobby, interest, and passion. Research in astronomy has been a source of fascination for him since the early 1980s. Since May 2002, Professor Filipović is affiliated with the Western Sydney University (WSU), and has been responsible for the development of Astronomy at WSU. Between 1997 and 2000, he held full-time research positions at the Max-Planck-Institut für extraterrestrische Physik (MPE; Germany). He is Chair of the largest public Observatory in Australia (the WSU's Penrith Observatory), and has over 200 refereed publications.

His research interests center on supernovae, high-energy astrophysics, planetary nebulae, Milky Way structure and mass extinctions, H ii regions, X-ray binaries, active galactic nuclei, deep fields (SPT & Pavo) and stellar content (WR, O, B stars) in nearby galaxies including the Magellanic Clouds. All of this research is closely related to further our understanding of the interactions between galaxies and the processes of stellar formation and star evolution as they affect galaxy evolution. One of the scientific highlights of Professor Filipović scientific career is flying on-board of NASA's Sofia science mission in June 2018. He is a fellow of the Astronomical Society of Australia, the Australian Institute of Physics, editorial board member of the Serbian Astronomical Journal and a member of the International Astronomical Union. He lives in Sydney with his family.

His ORCID is 0000-0002-4990-9288.

Nicholas F H Tothill

Nick Tothill was born in the south of England, read Physics at Corpus Christi College, Cambridge, and received his MSc from the University of Manchester, studying at Jodrell Bank. He graduated with a PhD in Astrophysics from the University of London, after research in star formation and the interstellar medium at Queen Mary & Westfield College. He has held research appointments in Germany, Canada, the United States, the United Kingdom, and Australia, and worked on telescopes in Spain, Chile, Hawaii, Arizona, Australia, and Antarctica. In 2004, he was a winterover scientist and Station Science Leader at the Amundsen–Scott South Pole Research Station.

In 2011, He joined Western Sydney University, where he is now Senior Lecturer in the School of Science and Director of the Penrith Observatory. He is a member of the Astronomical Society of Australia and the International Astronomical Union. His research still centers on the interstellar medium of the Milky Way, but includes topics as diverse as high-redshift galaxy surveys, Antarctic astronomy, and cosmic-ray astrophysics. He lives in Sydney with his family.

His ORCID is 0000-0002-9931-5162.

Foreword

It is my great pleasure to write this foreword for "*Principles of Multimessenger Astronomy.*" As a professional astronomer for more than two decades, and a past president of our national astronomy society, the Astronomical Society of Australia, I have seen a growing need for an astronomical text of this kind in our academic community. Having known the authors for many years, and having a great regard for their wide-ranging and deep knowledge spanning all aspects of astrophysics, it is no surprise to me that Miroslav and Nick have been the ones to bring this goal to fruition.

The authors are especially dedicated teachers as well as active researchers, and this book reflects their deep insight into the broad context of current astronomical research. Such expertize requires a dedicated understanding that not only encompasses the use of traditional astronomical techniques, spanning the entire electro-magnetic spectrum from radio wavelengths through the optical and infrared to X-rays and gamma-rays, but extends to the realms of quantum physics and general relativity in the forms of particle astrophysics and gravitational waves. This breadth of expertise is rare, and their overview of the field is timely and welcome!

This textbook provides an ideal introduction for those beginning a career in astrophysics. It brings together, in a single resource, the fundamental equations and principles of theoretical and observational astronomy, along with the relevant history to allow an understanding of its origins, and the necessary context for implementation. Its focus is broad, spanning the history of astronomy to the physics of the electromagnetic spectrum, particle astrophysics, and general relativity, as well as the practicalities associated with telescope development and their use in making measurements of the cosmos. It is unique in providing such an expansive overview. It recasts the astronomical research effort as a truly multimessenger exercise by bringing in particle astrophysics and gravitational wave astronomy as equally foundational alongside the more traditional electromagnetic approaches to under-standing the cosmos.

As might be expected of a volume covering such a far-reaching agenda, it delivers the key elements without being drawn into unnecessary detail. It supplements these overviews with a valuable set of references and suggestions for further reading to provide the necessary additional specifics for those interested in delving more deeply into any particular topic. Consequently the "*Principles of Multimessenger Astronomy*" serves as an excellent foundation for those new to astronomy, while also being a convenient and concise reference for active researchers. I hope you find as much of value in it as I have.

Andrew Hopkins,
Professor of Astronomy
Katoomba, Australia
2021 May 3

AAS | IOP Astronomy

Principles of Multimessenger Astronomy

Miroslav D Filipović and Nicholas F H Tothill

Chapter 1

Introduction to Multimessenger Astronomy: Fusing Intellect and Technology Through Telescopes

We introduce the four primary astrophysical messengers. These include electromagnetic photons, neutrinos, cosmic rays, and gravitational waves. This multimessenger approach to astronomy can best be explained by studying the history and the evolution of the detectors and telescopes used to study them. Once the basic principle has been established, it can be incrementally developed, refined, and improved in line with current technology.

1.1 Dream Big

"Dream big" was always the astronomers' motto—even in the early days of science. Scientific advances are, for the most part, incremental, building gradually on the work and results of others. Occasionally however, there is a "quantum shift" in our "world view," in our understanding of a particular phenomenon, or a theoretical breakthrough that changes everything we previously may have held dear. Sometimes such world views are hard to shift, but eventually even the most recalcitrant scientists are usually persuaded when the weight of experimental or observational evidence becomes overwhelming.

Often, such seismic shifts arise not through the power of intellect alone but by technology and the inventions that arise from combinations of mind and matter. One such breakthrough invention was the telescope. This single invention revolutionized our understanding of our world view of the Earth, solar system, Galaxy and the greater Cosmos. Let us not forget that Aristotle's[1] incorrect views persisted until

---

[1] Aristotle (384–322 Before the Common Era; BCE) was a Greek philosopher and polymath. His view was that the celestial realm was total perfection and all objects moved in perfect circles around the Earth with all celestial objects being perfect spheres (and perfect characteristics).

the time of telescopes. Technology can both drive, and be driven by, the curiosity of the astronomer to see further, deeper, fainter, and faster than ever before as new discoveries are uncovered.

The invention and use of the telescope, including its gradual refinement, also marked the birth of the modern scientific method. This revolution in scientific methodology led to a dramatic reassessment of our place in the universe and nature in general. The telescope, as a new technological device, showed conclusively that the universe is big—much bigger than assumed by our unaided senses. These discoveries opened a new chapter in human perception of the unforeseen vastness of the universe. Telescopes proved to us that our own Galaxy—the Milky Way—is only one among countless others. With every new generation of telescopes we are exposed to a wealth of new and exotic celestial objects and structures. From simple observers of the natural world around us, with the invention of the telescope we have grown into scientists that care about the real understanding of the processes in the Cosmos.

1.2 Four Astrophysical Messengers

The human eye has sensors that have the right properties to detect optical or so-called "visible" light. However, "invisible light" can be made "visible" with the right apparatus. Bees, for example, have the ability to see some ultraviolet light, and dogs and owls can see part of the infrared spectrum. In astronomy, while we have learned much about celestial objects using optical telescopes, there is much more to be discovered using other types of electromagnetic radiation and beyond (Figure 1.1). Multimessenger observation sees astrophysical processes and objects that are invisible to optical telescopes.

In the past century, modern observational astronomy has grown from the narrow wavelength band of optical light, which is about one octave[2] in width, to the entire electromagnetic spectrum and beyond to cosmic rays, gravitational waves and neutrino astronomy. Today, we routinely observe across more than sixty octaves, between the radio band and the highest energy gamma-ray radiation. In addition, new techniques to detect and measure cosmic rays, gravitational waves and neutrinos have provided new and complementary insights into celestial objects and events. These new observations and techniques paved the way for a new field of research called multimessenger astronomy.

> **Multimessenger astronomy is the careful analysis of the emissions by celestial objects of four messengers—photons (γ), neutrinos (ν), cosmic rays (p and other particles) and gravitational waves (Figure 1.2)—across all observable energy/wavelength ranges.**

To be more specific, multimessenger astrophysics is the experimental arm of modern astronomy, investigating astrophysical sources via multiple distinct messengers. Detection and subsequent detailed follow-up observations of such sources help us to unveil the underlying mechanism(s) that power them. These

---

[2] An octave is a frequency interval where one endpoint has twice the frequency, and half the wavelength, of the other endpoint.

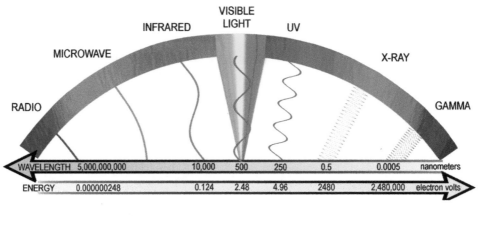

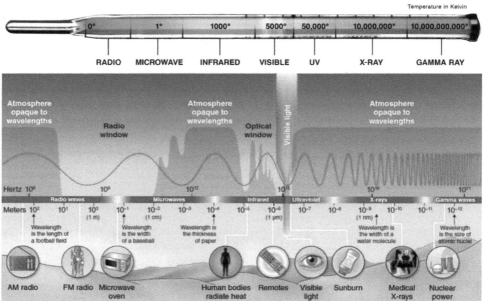

Figure 1.1. The electromagnetic spectrum, thermometer scale, and Earth's atmospheric opacity. Image credit: (top) NASA/CXC/S. Lee, (middle) NASA/CXC/M.Weiss, (bottom) NASA.

multimessenger observations also provide a unique opportunity to study the intrinsic properties of the messengers themselves (Cowen 2020). Within multimessenger astronomy, multifrequency (or multiwavelength) astronomy is a subset that deals exclusively with photons, but over a wide range of frequencies. Radio emissions are compared with optical and X-ray emissions, for example, to see if there is a correlation in position of these emissions and to determine the ratios of intensities of emissions both across and between these electromagnetic bands. These data are gathered from telescopes built on the surface of the Earth, from airborne (aircraft or balloon-borne) telescopes, and from telescopes in space (usually Earth or solar

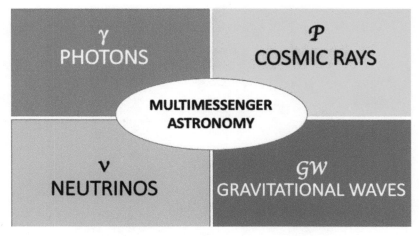

Figure 1.2. The four messengers of modern astronomy: photons (γ), neutrinos (ν), cosmic rays (p), and gravitational waves.

orbit). We need these non-ground-based telescopes because of the detrimental effects of the Earth's atmosphere on many observations (Section 1.4).

1.3 A Brief History of Telescopes: From 1608 to Neutrino and Gravitational Wave Telescopes

Way before the invention of the telescope, the stars were regarded as merely a convenient backdrop for scanning the wanderings of the Sun, Moon, and planets. Even the nature of these moving celestial wonders were unknown.

In modern astronomy, it is astrophysics that dominates, with its emphasis on understanding celestial objects and their natural properties. Pre-telescopic instruments such as the astrolabe, gnomon, merkhet, nocturnal, armillary sphere, cross staff, quadrant, dioptra, and others, were all made to estimate specific positions of celestial objects and its consequent motion. None of these instruments were designed to explain the universe. As pointed out by Robert Egler (2006), the transition from these instruments to telescopes fundamentally changed the nature of astronomy. Suddenly, we were far more interested in "what things are" than simply "where things are." Perhaps, a similar parallel could be drawn from our latest detection of three other astrophysical messengers that will again change and shift our understanding of the universe.

1.3.1 Optical Telescopes

Early Telescopes
The Invention of the Astronomical Telescope: The telescope emerged from the then-new Dutch optical industry—it is unclear precisely who invented it (though the credit is often given to Hans Lipperhey[3]; Watson 2004). By the end of 1608, reports of the invention were widespread. Galileo heard about it in 1609, and had designed

---

[3] Hans Lipperhey, often written Lippershey (1570–1619), German–Dutch spectacle-maker.

Figure 1.3. Replica of Galileo's first telescopes. Image credit: Shannon Stevenson.

his own, without seeing the original and incorporating improvements to increase the magnification to 20×, within days.

Galileo was the first modern scientist to build a telescope (Figure 1.3), point it at the sky, report his observations and discuss the implications. A patient and skillful observer, he accurately reported the cratered surfaces of the Moon, drew the Moon's phases in detail, described the Milky Way, discovered the rings of Saturn, and observed sunspots. He discovered Jupiter's moons, which he named the "Medicean stars"[4] (Figure 1.4). Galileo would quickly become convinced that Copernicus' Heliocentric model was absolutely correct.[5] His advocacy of this position, written in Latin and Italian, would ultimately see him put under house arrest by the Catholic Inquisition until his death in 1642 (McFadden 2018). Galileo understood three basic properties of a telescope:

- Light gathering power—the ability of a telescope to collect light,
- Resolving power—the ability of a telescope to reveal fine detail, and
- Magnifying power—the ability to make a resolved image appear larger.

Early Development—The 17th Century: The promise of the telescope for astronomical observations was immediately obvious, and the scientific community across Europe started to develop the field by developing new instruments and

---

[4] Today, we know them as the Galilean moons of Jupiter.
[5] Nicolaus Copernicus (1473–1543) was a Renaissance-era astronomer and mathematician. It is most likely that Copernicus developed his model independently of Aristarchus of Samos (310–230 BCE), another ancient Greek astronomer who had formulated such a model some 18 centuries earlier.

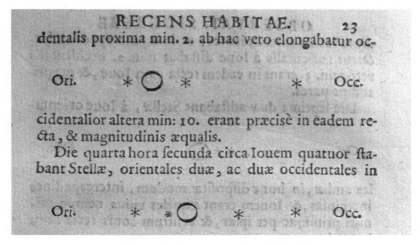

Figure 1.4. An excerpt from Galileo's Sidereus Nuncius published in March 1610, showing the "Medicean stars" (Io, Europa, Ganymede, and Callisto). This is the first published scientific work based on observations with an astronomical telescope. It also contains the imperfect Moon and stars of the Milky Way. Image credit: iau.org (History of Science Collections, University of Oklahoma Libraries) (CC BY 4.0).

making new observations. This combination of instrumental and scientific development, a hallmark of astronomy, continues to this day.

Although Johannes Kepler[6] used pre-telescopic data to formulate his laws of planetary motion, he made detailed studies of optics and devised a telescope design with two convex lenses—the "Keplerian Telescope." His pioneering book "Astronomia Pars Optica" (Figure 1.5) would earn him the title of "Founder of Modern Optics" (McFadden 2018). Compared to the Galilean telescope, the Keplerian design gives improved magnification and produces a real image, which in turn allows for measurement devices such as micrometers to be placed in the image plane. However, the Keplerian design produces inverted images—a drawback that astronomers can easily tolerate.

As the telescope became widely adopted, it was realized that larger telescopes generally performed better. In particular, long telescopes with large focal lengths suffered less from chromatic aberration, and thus had clearer images.

Astronomers such as Huygens[7] and Hevelius[8] led the move toward larger and larger telescopes over the rest of the 17th century. Huygens used these larger telescopes to observe Saturn and its rings, showing that they were truly rings, and not "ears," and discovering Titan, Saturn's largest moon.[9]

The First Reflecting Telescopes: Because curved mirrors had long been known to have optical effects similar to those of lenses, the possibility of using them as optical

---

[6] Johannes Kepler (1571–1630), German astronomer, mathematician and astrologer.
[7] Christian Huygens (1629–1695), Dutch astronomer.
[8] Johannes Hevelius (1611–1687), Polish astronomer.
[9] The European Space Agency's Cassini mission to Saturn carried a probe called Huygens that landed on Titan.

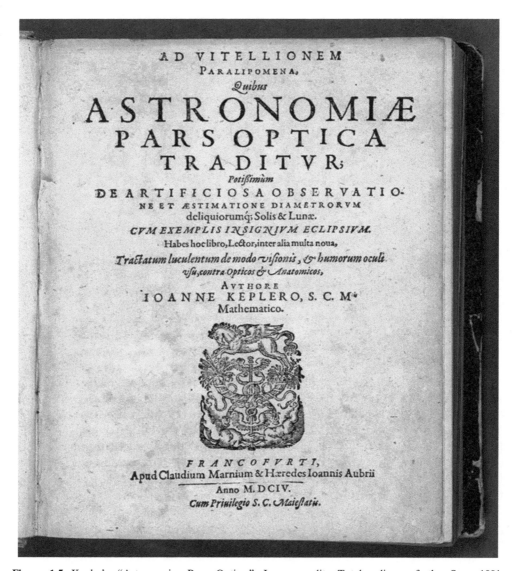

Figure 1.5. Kepler's "Astronomiæ Pars Optica." Image credit: Total eclipse of the Sun, 1881, priJLC_SCI_002968, Jay T. Last Collection of Graphic Arts and Social History, The Huntington Library, San Marino, California.

elements in telescopes was investigated soon after the invention of the telescope. Although James Gregory[10] published a detailed design in his "Optica Promota" of 1663, the first working model was Isaac Newton's reflector of 1668. Robert Hooke (1635–1703) built the first *Gregorian* telescope in 1673. Both Newtonian and Gregorian optical designs are still in use today.

---

[10] James Gregory (1638–1675), Scottish mathematician and astronomer.

Reflectors offered substantial advantages over refracting telescopes: a mirror reflects all wavelengths of light in the same way, and so there is no chromatic aberration, which was the limiting factor on early refractors. The telescope was more compact than a refractor, and comparatively easy to build. However, the metal mirror surfaces used at the time, had low reflectivity, and thus lost a lot of light.

18th Century Telescopes: Both reflectors and refractors were extensively used in the 18th century, but the development of achromatic optics for refractors made them the instrument of choice for many astronomers. Isaac Newton had considered chromatic aberration to be an insoluble problem, and this led him to design a reflecting telescope. Achromatic optics, however, by combining different types of glass that compensated for each other's chromatic aberration, greatly reduced the problem, so that the refractor remained a useful instrument. Chester More Hall[11] designed achromatic lenses and used them to construct the first achromatic refracting telescope around 1730, but did not publish his design. The optician John Dollond[12] patented achromatic lenses, or achromats, in 1758 and went on to construct telescopes using them.[13] The original achromats combined crown glass (with low refractive index, containing potassium oxide) and flint glass (with high refractive index, containing lead oxide).

Reflectors were improved by the introduction of the Cassegrain design[14] and by greatly improved mirror shapes due to John Hadley.[15] William Herschel[16] pushed the technical development of reflecting telescopes in order to advance his scientific agenda, and most of his work was done with his own reflectors, using his own mirrors. Although he built a 1.2 m aperture telescope (Figure 1.6), its results were disappointing, and his earlier 0.5 m aperture telescope was his workhorse. This 0.5 m (~20 inch) instrument represents the summit of reflecting telescope technology around the start of the 18th century.

The balance between reflector and refractor can be seen in the instruments used by Charles Messier[17] (a contemporary of Herschel's). His favorite instrument seems to have been a Gregorian reflector of about 0.2 m aperture, later moving to a smaller (~0.1 m) achromatic refractor. The smaller refractor had similar effective aperture to the larger reflector, simply due to the limitations of the speculum metal mirror used in the reflector.

Speculum metal is an alloy of copper and tin (about 30% tin, compared to the ~12% tin in bronze), that can be polished to a highly reflective surface. However, this surface tarnishes in humid conditions, losing reflectivity, and requiring constant

---

[11] Chester Moore Hall (1703–1771), British lawyer and inventor.

[12] John Dollond (1706–1761), English optician.

[13] It is still unclear whether Dollond copied Hall's work, or invented achromats independently.

[14] Laurent Cassegrain (1629–1693), Catholic priest and astronomer.

[15] John Hadley (1682–1744), English mathematician and inventor of the octant.

[16] William Herschel (1738–1822), German-born British astronomer and composer, discoverer of Uranus.

[17] Charles Messier (1730–1817), French astronomer.

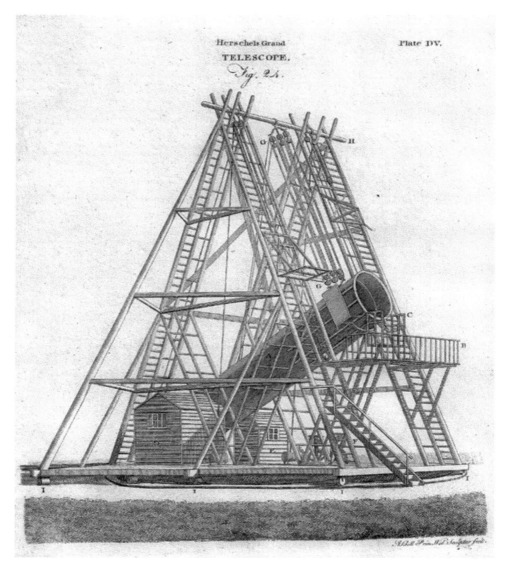

Figure 1.6. An illustration from the 1797 edition Encyclopedia Britannica Volume 18 of William Herschel's 1.2 m (48 inch) reflecting telescope with 12 m (40 foot) focal length, also called his "40 foot telescope." Image credit: Andrew Bell and Colin Macfarquhar, Wikimedia Commons.

polishing. This polishing alters the shape of the metal, and the mirror therefore needs to be reground. Herschel's reflectors were able to out-compete refractors of the time because he could make, polish and regrind mirrors large enough to overcome the reflector's weakness as an instrument.[18]

---

[18] Another element in his success was removal of the secondary mirror—an extra source of light loss—to give the "Herschelian" optical design.

19th Century: Rise of the Giant Telescopes

At the end of the 18th century, achromatic lenses for refracting telescopes were limited to diameters of about 0.1 m (4 inches). But by 1800, Guinand[19] had learned how to make larger flint glass lenses, and he then taught the technique to Fraunhofer.[20] In the first quarter of the 19th century, Fraunhofer supplied refracting telescopes with apertures up to about 10 inches (0.25 m). His immediate successors supplied 15 inch (0.4 m) telescopes to Russia and the USA.

Refractors made by telescope makers such as Thomas Cooke in England, Sir Howard Grubb (1844–1931) in Ireland, Alvan Clark (1804–1887) in Massachusetts, and John Brashear (1840–1920) in Pennsylvania were the observatory workhorses of the 19th century. Ten inch apertures (0.25 m) were common, while 20–40 inch aperture refractors produced landmark discoveries. E. E. Barnard (1857–1923) discovered the fifth moon of Jupiter in 1892[21] using the Lick 91 cm (36 inch) refractor. Also, Pluto was discovered in 1930 with the Lowell Observatory 33 cm refractor. According to Agnes Mary Clerke[22] in 1887: "Refractors have always been found better suited than reflectors to the ordinary work of observatories as they lend themselves with far greater facility to purposes of exact measurement."

The culmination of refractor development was reached with the 1 m (40 inch) telescope at Yerkes Observatory, Wisconsin, which was completed in 1897 and housed in a 27.5 m diameter dome. Alvan Clark,[23] who made this 1 m lens, felt that a still larger refractor would be impossible because the weight of the glass would cause the lens to sag; unlike a mirror, a lens can only be supported around its circumference (Kramer 2012).

The 1.25 m Great Paris Exposition telescope was the largest refractor ever built, with a very unconventional layout: the telescope was mounted horizontally, observing the sky by means of a siderostat. The telescope performance was poor, and it was very impractical. After the Exposition, it was broken up for scrap (Kramer 2012), while slightly smaller observatory instruments, such as the Yerkes and Lick refractors, continued to contribute to astronomy.

These giant refractors were very expensive instruments. According to William R. Harper[24] in 1892: "The whole enterprise will cost Mr Yerkes certainly half a million dollars. He is red hot and does not hesitate. It is a pleasure to do business with such a man" (Osterbrock 2002).

Despite the limitations of speculum mirrors, reflectors continued to develop during the 19th century. The biggest speculum-mirror reflector was the "Leviathan of Parsonstown," a 1.8 m (6 foot) aperture instrument, completed in 1840 by the Earl of Rosse.[25] The development that truly allowed reflectors to become the

---

[19] Pierre Louis Guinand (1748–1824), Swiss optician.

[20] Joseph von Fraunhofer (1787–1826), German physicist and optician.

[21] Expanding beyond the four moons originally found by Galileo.

[22] Agnes Mary Clerke (1842–1907) was an Irish astronomer and writer, mainly in the field of astronomy.

[23] Alvan Clark (1804–1887), American astronomer and telescope maker.

[24] William Rainey Harper (1856–1906), American academic leader and Baptist clergyman.

[25] William Parsons, 3rd Earl of Rosse (1800–1867), Anglo-Irish astronomer.

dominant telescope type came in 1857, when Steinheil[26] and Foucault[27] invented a process to deposit a silver coating on glass and used it to produce telescopes with silvered-glass "first-surface" mirrors. These mirrors were highly reflective, and could be recoated without loss of the mirror shape. The last major speculum-mirror reflector was the 1.2 m Great Melbourne Telescope.[28]

20th Century Modern Telescopes

The 20th century brought the age of refractors to an end. The ones that were built were still used extensively, but there was no further development of them as flagship instruments. That mantle was taken over by the reflectors. A. A. Common's[29] series of glass-mirror reflectors (0.5, 0.9, and eventually 1.5 m apertures) showed the potential of the technology at the end of the 19th century, and were the last world-leading instruments to be constructed and used by an amateur. At the start of the 20th century, the newest telescopes were silvered-glass reflectors, mainly sited outside Europe.

The development of larger and larger reflectors was driven by George Hale.[30] Following his establishment of the Yerkes Observatory, with its 1 m refractor, Hale founded the Mount Wilson Observatory outside Los Angeles with a 1.5 m reflector completed in 1908. In 1917, this was joined by the 2.5 m Hooker telescope,[31] which was the largest-aperture telescope in the world until 1949. Hubble used this telescope to establish the expansion of the universe, and Michelson used it to measure the diameter of stars by interferometry. Hale then started to work toward the establishment of the Palomar Observatory, also in southern California. The 5.1 m (200 inch) Hale telescope was completed in 1949, after Hale's death, at Palomar, and was the largest in the world for a quarter century.

New developments in mirror technology were crucial to this evolution. In 1930, John Donovan Strong (1905–1992) invented the process of "aluminizing" in which an aluminum coating was applied to the glass mirror by vacuum deposition, resulting in a more reflective and longer-lasting surface. This technology was applied to the Mount Wilson telescopes in the 1930 s. The 5 m Hale telescope combined this with the use of low-expansion Pyrex glass for the mirror, reducing thermal effects on the telescope over the day/night cycle.

Past the 5 m Aperture Limit: Hale's telescopes set the pattern for the large astronomical telescopes of the mid-20th century: an aluminized mirror made of a low-expansion borosilicate glass (such as Pyrex) or glass-ceramic (such as Cervit, Zerodur) on an equatorial mounting. This basic model can be seen in the 3 m Shane telescope (Lick Observatory, California, 1959), the 3.5 m Calar Alto telescope (Spain, 1984), the European Southern Observatory (ESO) 3.6 m telescope (La

---

[26] Carl August von Steinheil (1801–1870), German physicist and astronomer.
[27] Léon Foucault (1819–1868), French physicist.
[28] Later moved to Canberra, and destroyed by the Mount Stromlo bushfire in 2003.
[29] Andrew A. Common (1841–1903), British amateur astronomer and optical designer.
[30] George Ellery Hale (1868–1938), American astrophysicist.
[31] The mirror blank for the Hooker telescope was cast out of green glass by the St. Gobain Glass Works in France.

Silla, Chile, 1977), the 3.6 m Canada–France–Hawaii Telescope (Maunakea, Hawaii, 1979), the 3.9 m Anglo-Australian Telescope (Siding Spring, Australia, 1974), the 4 m Mayall and Blanco telescopes (Kitt Peak, Arizona, 1973, and Cerro Tololo, Chile, 1976, respectively).

From this list, it is clear that the 5 m Hale telescope remained the practical limit for telescope aperture. The 6 m BTA-6 (Bolshoi Teleskop Alt-azimutalnyi), built by the then USSR in 1975, surpassed the Hale Telescope, but was prevented from reaching its potential by technical flaws. The BTA-6 did, however, reintroduce the use of alt-azimuth mounts to large-telescope astronomy. Much simpler to engineer than equatorial mounts, alt-azimuth mounting systems were used by William Herschel for his large telescopes, and by Lord Rosse for the Leviathan of Parsonstown. Equatorial mounts, with their superior celestial tracking, had decisive advantages for the long observations required by the 20th century disciplines of photography and spectroscopy. The BTA-6 used computer control to compensate for the alt-azimuth system's drawbacks, and computerized alt-azimuth mounts were used for essentially all subsequent large telescopes, starting with the 4.2 m William Herschel Telescope (La Palma, Canary Islands, 1987). Even within the 4 m class, alt-azimuth mountings became ubiquitous: the 3.6 m ESO NTT (La Silla, Chile, 1989), ARC 3.5 m telescope (Apache Point, New Mexico, 1994), WIYN 3.5 m telescope (Kitt Peak, Arizona, 1994), 3.6 m Telescopio Nazionale Galilei (La Palma, 1997), and 4.1 m SOAR telescope (Cerro Pachón, Chile, 2002) all use alt-azimuth mounts.

Alt-azimuth mounting systems, with their simplified and cheaper mounting and enclosure requirements, came together with further innovation in mirror-building to allow the 5 m mark to be passed at the end of the 20th century. Roger Angel (1941–) constructed thin lightweight mirrors with honeycomb structures, using spin-casting[32] to produce a parabolic shape before any grinding or polishing, in the University of Arizona's Mirror Lab. The 6.5 m mirror blanks made at the Mirror Lab were used to refit the Multiple Mirror Telescope (see below) and build the twin Magellan Telescopes (Las Campanas, Chile, 2000–2002). Larger monolithic mirrors were cast by Corning for the twin 8.1 m Gemini telescopes (Maunakea & Cerro Pachón, 1999–2001) and the 8.2 m Subaru telescope (Maunakea, 1999). The 8.2 m mirrors for the four telescopes of the ESO Very Large Telescope (VLT; Paranal, Chile, 1998–2001), were cast in Zerodur by Schott and polished by REOSC (https://www.safran-electronics-defense.com/). These 8 m class telescopes built at the turn of the 21st century represent the current limit for the single-mirror reflecting telescope as developed by Newton and Gregory.

21st Century—The Path to an Extremely Large Telescope
The first large telescope to use multiple mirrors to substitute for a single larger mirror was the Multiple Mirror Telescope (Mount Hopkins, Arizona, USA 1979), which brought the light gathered by six 1.8 m mirrors to a common focus, yielding

---

[32] This is a technique for making large parabolic mirrors by using the curved surface created by a rotating liquid.

an effective aperture of 4.5 m. These mirrors were replaced by the first of the Mirror Lab's 6.5 m mirrors in 1998.

A slightly different approach was taken by the Large Binocular Telescope (LBT; Mount Graham, Arizona, USA, 2005). The LBT consists of two telescopes on a common mount, rather than two mirrors with a common focus. Each primary mirror is 8.4 m across, making them the largest monolithic mirrors in the world. Although the two telescopes have their own secondary mirrors and other optics, they are so closely aligned that they can be used as an interferometer.

The largest telescopes now use segmented mirrors, in which a large number of hexagonal segments are mounted together, but can be individually adjusted to maintain the overall shape of the mirror. This technology was developed by Jerry Nelson (1944–2017) for the twin 10 m Keck telescopes (Figure 1.7; Maunakea, USA, 1993–1996). The 10.4 m Gran Telescopio Canarias (La Palma, Spain, 2007) also uses a segmented primary. Segmented mirrors are key elements of the innovative design of the 10 m Hobby–Eberly Telescope (Mount Fowlkes, Texas, USA, 1996) and the very similar South African Large Telescope (Sutherland, South Africa, 2005). These telescopes' segmented primaries are spherical instead of paraboloidal, and they rotate in azimuth only, the primary mirrors having fixed altitudes. This unusual design sacrifices some elements of telescope performance, but

Figure 1.7. The twin Keck telescopes on the summit of Maunakea, Hawaii. The Subaru telescope is to the left, and the NASA Infrared Telescope Facility to the right.

retains those necessary for the intended programs, at a greatly reduced cost compared to a more general-purpose 10 m telescope. The 5 m LAMOST telescope (Xinglong, China, 2008) is even more specialized than HET and SALT, using segmented optics solely to deliver multiobject spectroscopy.

A new generation of *Extremely Large Telescopes (ELTs)* is under development and construction. These instruments have apertures of around 30 m. The Giant Magellan Telescope (GMT; Las Campanas, Chile) has an optical layout similar to the original MMT design, with six 8.4 m monolithic mirrors produced by the Mirror Lab to give an overall aperture of 25 m. The Thirty Meter Telescope (TMT; Maunakea, USA) and 39 m European Extremely Large Telescope (E-ELT; Cerro Armazones, Chile) both use segmented primary mirrors. There was even a design study for a 100 m Overwhelmingly Large Telescope (OWL); the reasons for abandoning this study were budgetary rather than technical, suggesting that the segmented mirror, alt-azimuth telescope model may have room to grow past even the ELTs (Figure 1.8).

From this historical overview of ground-based optical telescope development over the past 400 years, one can clearly conclude that the most important change in telescopes over time is their aperture size. The principle is simple—larger telescopes collect more light and so we can see fainter objects. Between Galileo's telescope and the Keck telescopes, diameters have grown by over a factor of 600, while their areas have grown by over 400,000 times. Figure 1.9 shows a form of Moore's law for optical telescopes. According to this relationship, the light collecting area of the largest telescopes has doubled every (approximately) 20 years over the 400 year history of telescopes. The trend is not smooth as there can be a large jump in the size of the largest mirror as a new technology becomes available.

The Search for Better Locations
As telescopes and our understanding of astronomical observations have evolved, it has become clear that ideal locations for astronomy tend not to be favored locations for human activity—and, indeed, that human activity can significantly damage the quality of a potential observing site. There is, therefore, a balance between convenience (which in turn affects logistics, cost, and human factors) and absolute quality of an observing site. Over the centuries, this balance has generally shifted more and more toward site quality. In the 21st century, locations as inhospitable as the high Andes mountains, the Antarctic Ice Cap, and space (even beyond most satellite orbits) have become favored locations.

Local Observatories: Early telescopes were simply located in their users' houses, gardens, or estates. Larger observatories were founded in or near cities—Paris Observatory in 1667, and the Royal Greenwich Observatory near London in 1675. By the late 19th century, although these urban observatories were widespread, it was clear that new sites were needed, away from the lights and smoke of cities. The Yerkes Observatory, for example, was founded in 1897 some 100 km from Chicago, about 300 m above sea level.

Mountain Observatories: The move from sea-level observatories to mountain observatories coincided with the shift to large reflectors at the end of the 19th

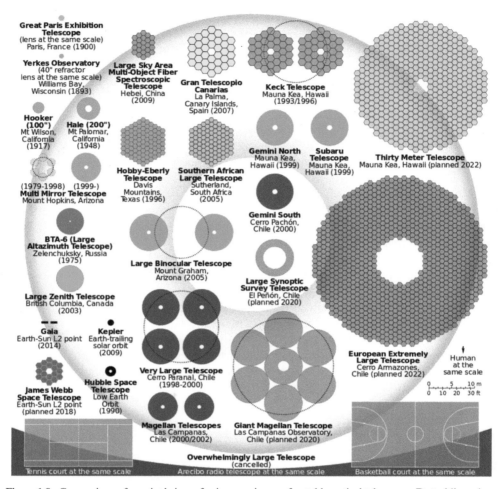

Figure 1.8. Comparison of nominal sizes of primary mirrors of notable optical telescopes. Dotted lines show mirrors with equivalent light gathering ability. Image credit: Wikipedia: Cmglee (CC BY-SA 3.0).

century, and was also dominated by the new American astronomy establishment. Experiments with high-altitude observations were carried out in the late 19th century. Jules Janssen[33] realized that atmospheric features could be seen in astronomical spectroscopy, and, reasoning that higher altitude and less air would reduce the effect, placed instruments on mountain-tops, even Mont Blanc.

The first permanent mountain observatory was the Lick Observatory (Mt Hamilton, California, 1300 m altitude, founded 1888, preceding even the low-elevation Yerkes observatory). Around the same time, the Pic du Midi Observatory (France, 2900 m altitude) and Lowell Observatory (Flagstaff, Arizona, 2200 m

---

[33] Pierre Jules César Janssen (1824–1907) was the co-discoverer with J. Norman Lockyer of helium in the solar spectrum.

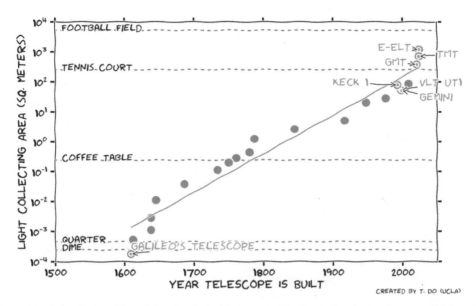

Figure 1.9. An analogy to Moore's law for Optical Telescopes. The line in the plot is an exponential fit to the relationship between the time a telescope was completed and its collecting area $A = A_0 \exp(t/\tau)$ (area = A_0 exp (time/tau)). Image credit: T. Do (UCLA; https://www.followthesheep.com/?p=1322).

altitude, 1894) were founded. Subsequently, Hale's Californian observatories were founded on the 1700 m peaks of Mount Wilson and Palomar. Astronomical observatories proliferated across the southwestern USA over the rest of the 20th century. This development centered on Arizona, starting with the 2100 m Kitt Peak (1960) and taking in Mount Hopkins (2600 m, 1966) and Mount Graham (3200 m, 1993), but also occurred in the neighboring state of New Mexico (Sacramento Peak, 2800 m, 1984; Magdalena Ridge, 3200 m, 1999).

The most important development in the USA in the second half of the 20th century was the development of the Maunakea observatory site (Hawaii, 4200 m, 1968). The summit of a volcano in the Pacific Ocean, this site quickly became one of the most important in the world, hosting 4 and 10 m class telescopes. Maunakea will also host the Thirty Meter Telescope, currently being developed. The Haleakala observatory at the summit of Maui (Hawaii, 3000 m, 1964) is less significant, but still well-used.

A few mountain observatories were built in Europe over the 20th century, such as the Calar Alto observatory in Spain (2200 m, 1975), but in general, European mountain sites were found to have quite poor atmospheric conditions. European astronomers therefore became leaders in siting their observatories outside their home countries. The UK collaborated with Australia to develop telescopes at the Siding Spring Observatory (New South Wales, 1200 m, 1962). The main focus of this work, however, was in the Andes mountains of South America, where the ESO founded the La Silla Observatory (2400 m) in 1969, shortly after US-based institutions created the Cerro Tololo Inter-American Observatory (Chile, 2200 m,

1965). Chile became the major location for telescopes in the Southern hemisphere (as Maunakea was in the North), with new observatories at Cerro Las Campanas (2400 m, 1971), Cerro Pachón (2700 m, 2000), and Cerro Paranal (2600 m, 1999). Two of the three ELTs under development will be sited in Chile—the Giant Magellan Telescope at Las Campanas, and the E-ELT at Cerro Armazones (2800 m, 1995).

The success of this model—that of constructing and operating telescopes at the best site available, rather than the best site in the telescope's home country—led to the development of the Canary Island observatories as well. The Canaries are a chain of volcanoes in the Atlantic Ocean, comparable to the Hawaiian island chain in the Pacific. The Roque de los Muchachos Observatory (ORM) was established at 2400 m around 1985. Although the Teide observatory (on a different island) had already been operating at similar altitude for about 20 years, the ORM became the site for the 4 m class William Herschel Telescope (WHT) and 10 m class Gran Telescopio Canarias (GTC).

Observatories in the 21st Century: In the 21st century, world-class telescopes (such as the ELTs) are such large-scale projects that to place them anywhere other than the best observing site is a gigantic waste of resources. We have arrived at the opposite approach to the domestic telescopes of the early days; we bring the resources to the telescope, not the other way around.

The investments in ELT projects are also so large that knowledge of the site characteristics cannot be left to chance. The advances made in the late 20th century in understanding the atmosphere above observatories and its effects on astronomical observations were integrated into large-scale 21st century site-testing campaigns that contributed significantly to the final site selection for these telescopes.

Recently, there has been a move toward evaluating new observatory sites in the high-altitude areas of Asia, driven largely by the increased capacity of nations such as China and India. The Indian Astronomical Observatory has been operating in the Himalayas since the turn of the century, at an altitude of 4500 m, while the Chinese Ali Observatory in Tibet was founded at 5100 m in 2011.

Polar Observatories: In contrast to the mountainous terrain usually used to get telescopes to altitude, the Earth's polar ice caps are very high and very flat. In addition to their altitude, the polar regions enjoy long periods of uninterrupted darkness, without the diurnal cycle of other parts of the planet. Intensive site testing has shown that atmospheric turbulence is confined to a thin boundary layer above the ice surface, and small telescopes have already been deployed to several sites on the Antarctic plateau. Testing has also been undertaken on Ellesmere Island in the Arctic.

Airborne Telescopes: Using aircraft of some sort to get a telescope even higher than a mountaintop is an obvious extension to the mountain observatory paradigm. The practical restrictions on size and weight of instrumentation place significant hurdles in the way, however, and the benefit expected from airborne observing in visible light seems not to be worth the trouble. Airborne astronomy has really found its niche in wave bands outside visible light.

Space Telescopes: The final step in escaping the effects of the atmosphere is to leave it behind altogether, by placing a telescope into space. While space-based astronomy is dominated by those wave bands to which the atmosphere is opaque (e.g., IR, X-ray), there are also considerable advantages in putting an optical telescope in space. These advantages accrue both from the elimination of atmospheric absorption and from the elimination of atmospheric turbulence (Section 5.2.2). Atmospheric turbulence limits the spatial resolution of optical telescopes at the best mountain sites to about half an arcsecond (depending on site conditions), much coarser than their aperture sizes would suggest. The turbulent terrestrial atmosphere not only blurs the light passing through it, degrading spatial resolution, it also introduces small variations in the transmission of light, degrading our measurements of brightness. The measured brightness of a star will vary randomly with time (*photometric noise*), potentially masking real variations of astrophysical interest.

The Hubble Space Telescope (HST; Figure 1.10) is the only general-purpose optical observatory in space (as opposed to special-purpose instruments). Its chief advantage over ground-based telescopes is that it has the full resolution implied by its 2.4 m aperture (Figure 1.11). Also, the lack of atmospheric absorption has also contributed to its mission by allowing it to observe ultraviolet and infrared radiation. As an "orbiting observatory," HST accepts observing proposals from the astronomical community, rather than carrying out a single observing program. It carries a variety of imaging and spectroscopic instruments to carry out these observations, and—very unusually—these instruments can be exchanged by

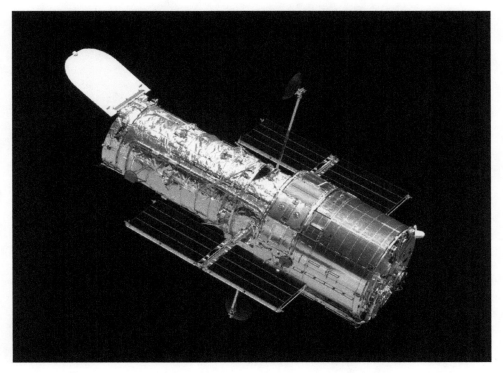

Figure 1.10. The Hubble Space Telescope (HST). Image credit: Ed Summers, Matthew Horn SAS/NASA.

Ground: Subaru (8m)   Space: *HST* (2.4m)

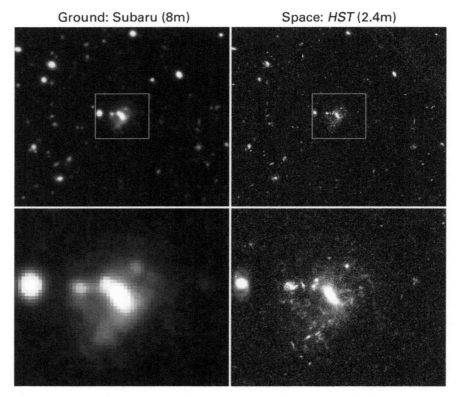

Figure 1.11. Hubble Ultra Deep Field (HUDF). A comparison of imaging capabilities from the 8 m Subaru telescope on Maunakea and the 2.4 m HST using the ACS (Advanced Camera for Surveys) of the GOODS (Great Observatories Origins Deep Survey) North field. The Subaru image was taken when seeing was about 0.8" though it can get down to 0.4" under exceptional conditions on Maunakea. The resolution for the HST image is 0.08". Image credit: NASA, Mauro Giavalisco, Lexi Moustakas, Peter Capak, Len Cowie, and the GOODS Team.

astronauts carrying out "service missions." No more service missions are planned due to the retirement of NASA's Space Shuttle fleet, so HST is now in its final observing configuration. Nevertheless, after launching to orbit in 1990, its service life of three decades demonstrates its usefulness; and is also an example of the principle seen on the ground—that constant development and replacement of instruments can greatly extend the useful lifetime of a telescope.

An optical telescope in space is also immune to photometric noise from the atmosphere, as demonstrated by the repurposing of the star-tracking guidance telescopes on the defunct WIRE satellite around 2000. These results were sufficiently promising to give rise to a series of optical satellite telescopes designed purely for this purpose: The Canadian MOST and the European CoRoT missions concentrated on asteroseismology, looking for stellar pulsations. CoRoT was also capable of detecting extrasolar planets by observing their transits of their parent stars. The Kepler mission and its successor Transiting Exoplanet Survey Satellite (TESS) are both specialized exoplanet-hunters.

1.3.2 Infrared Telescopes and Observatories

William Herschel's best-known contribution to astronomy may have been the discovery of Uranus, but he made an equally important contribution to astrophysics by discovering infrared light in 1800. He found that heat from the Sun could be measured without any corresponding light, and that this "radiant heat" could be refracted, behaving in the same way as visible light.

Infrared light cannot be experienced by humans in the same way as visible light; its presence and brightness must be inferred by measuring its effects on a physical system. Herschel's detector was the bulb of a mercury thermometer which he could then read. Later in the 19th century, devices were developed whose electrical properties changed with temperature. These electrical measurements were both accurate and sensitive, so the devices could be used to measure the small changes in temperature induced by infrared radiation.

In 1856, Charles Piazzi Smyth (1819–1900) used thermocouple detectors in Tenerife.[34] In these detectors, a small change in the temperature of two dissimilar metals makes a current flow, and they were also used by Lord Rosse to detect infrared light from the Moon with his "Leviathan of Parsonstown" telescope in 1870.

The bolometer, developed by Langley[35] around 1880, used the temperature dependence of electrical resistance to detect the heating effect of infrared light. The key to its success was its use of an electrical circuit called a Wheatstone bridge to measure the resistance very accurately. Until the mid-20th century, infrared astronomy was a very limited field, using thermocouples and bolometers to observe solar system objects and the brightest stars only.

The first infrared-sensitive photoconductor, lead sulfide (PbS), was developed as a detector during the Second World War and was applied to infrared astronomy in the 1950s. In a photoconductor, individual photons promote electrons in the material into the *conduction band*. This microscopic process can be detected through its macroscopic effects: the generation of a photocurrent or a reduction in resistance. Photoconductor detectors are thus qualitatively different to thermocouples and bolometers, since they detect the arrival of photons rather than the general heating effect of radiation impinging on the detector element.

Liquid nitrogen-cooled PbS detectors were used by the Two-Micron Sky Survey team at Mount Wilson to detect thousands of sources of infrared radiation. Because it did not target known astronomical objects, but rather scanned the sky, this survey succeeded in discovering astronomical objects with no known visible-light emission.

In the 1980s, arrays of photoconductor cells were used to build *infrared cameras*, using more sensitive photoconductors such as indium antimonide (InSb) and

---

[34] One of the Canary Islands—the Teide site that he used is still an observatory today, though less prominent than La Palma.
[35] Samuel Pierpont Langley (1834–1906), American astronomer and physicist.

mercury–cadmium–telluride (HgCdTe). These and other infrared instruments were often deployed to specialized infrared telescopes such as IRTF (NASA Infrared Telescope Facility) and UKIRT (UK Infrared Telescope), both on Maunakea, but, over time, were often used with visible-light telescopes as well.

This has led to the current situation, in which ground-based optical astronomy is often called *optical–IR (OIR)* astronomy. Telescopes are optimized for both visible and near-infrared light, and infrared instrumentation is a vital component of their instrument suites (see Appendix F). Indeed, the near-infrared is sometimes considered to be the core wave band for modern telescopes, for both astrophysical and technological reasons.

The development of the germanium bolometer in the 1960s was a similar leap in capability. The combination of the high temperature-sensitivity of germanium's resistance and cryogenic cooling gave much better sensitivity. While these bolometer detectors were used on ground-based telescopes, their most immediate impact came in airborne astronomy.

Airborne Infrared Astronomy
In 1856, Pizzi Smyth (1819–1900) found that his infrared signals got stronger as he moved to higher elevations, showing that the atmosphere was attenuating the infrared radiation. At longer infrared wavelengths, an additional problem is presented, as the atmosphere itself is warm enough to emit radiation at the wavelength of interest, so that a celestial object is seen as a background object to a foreground screen of emission. Clearly, the more atmosphere one could rise above, the less attenuation there will be. In addition, there will be less atmosphere above the observer, and that atmosphere will be colder. Both of these act to reduce the problems of infrared observation (see Chapter 5).

In 1968, a 0.3 m infrared telescope mounted in a NASA Learjet was a successful proof of concept, and a more permanent airborne observatory was developed. The Kuiper Airborne Observatory (KAO) fielded a 1 m infrared telescope aboard a dedicated Lockheed C-141 aircraft, flying observing missions in both northern and southern hemispheres (based at NASA Ames Reseach Centre in California and in Australasia respectively). The KAO was retired to make way for the German–American Stratospheric Observatory For Infrared Astronomy (SOFIA) project, with a 2.5 m telescope in a specially converted Boeing 747 aircraft (Figure 1.12).

All of these airborne missions fly rather higher than normal air passenger traffic, around 12–13 km, or ~43,000 feet. They represent a compromise between ground-based and space-based instrumentation. Airborne observatories have to contend with more atmospheric absorption than space observatories, but much less than even the highest mountaintop observatories (the highest of which is about 6 km). The constraints of cost, size, and the time required to develop them are likewise somewhere between those of ground-based and space-based telescopes. Although the airborne telescope itself cannot be significantly altered once built, new instruments can be developed for it to extend its capabilities past its original design—an option generally unavailable in space.

Figure 1.12. Stratospheric Observatory For Infrared Astronomy—SOFIA. The open shutter behind the wing allows the 2.5 m telescope to see out. Image credit: NASA.

IR Astronomy from Space

The best way to avoid the problems of atmospheric absorption and emission is by observing from space. Space-based infrared observations are almost as old as spaceflight. In the 1960s and 1970s, Project Hi Star at the US Air Force Cambridge Research Lab (later known as the Geophysical Lab)[36] flew infrared detectors on suborbital rockets, which had a few minutes' observing time before re-entry. They created the first infrared all sky map, consisting of data from a total of 30 minutes observation time. Their observations at wavelengths of 4, 10, and 20 microns detected 2363 reliable infrared sources, about 70% of which could be matched to sources in the Mount Wilson 2.2 micron survey.

Balloon-borne infrared telescopes fall somewhere between airborne and space observatories. The weight limits on balloon payloads are stringent, but balloons deliver much longer observing times than suborbital rockets. They are able to ascend to higher altitudes than aircraft (~40 km), so they are most useful at the longer wavelengths, where the high altitude makes observations possible. For example, a

---

[36] Many infrared sources are still known by designations starting with CRL or AFGL.

balloon-borne bolometer experiment surveyed the sky at a wavelength of 100 microns in the 1960 s, discovering 120 bright sources near the plane of our Galaxy.

Orbital IR Telescopes: An infrared telescope in orbit, however, has all the advantages of space, while enjoying a long duration. Most infrared telescopes and detectors require cryogenic cooling, which is most easily achieved by launching with a supply of liquid cryogen, usually liquid helium. The limits of this supply often define the duration of observations.

The first orbital infrared telescope, the Infrared Astronomy Satellite (IRAS) was a collaboration between the US, UK, and Netherlands, observing for 10 months in 1983 before its cryogen supply ran out. With a 0.6 m aperture telescope, IRAS surveyed 96% of the sky at wavelengths of 12, 25, 60, and 100 microns, producing maps and catalogs (of over 350,000 sources) that were the fundamental references in this wave band for decades, and discovering dust disks around stars (such as Vega) and infrared-bright galaxies.

In July and August 1985, infrared observations on-board the Space Shuttle's Spacelab 2 complemented the observations made by the Infrared Astronomical Satellite (IRAS) mission, producing a high quality map of about 60% of the Galactic Plane.

The Cosmic Background Explorer (COBE), launched by NASA in late 1989 was mainly intended to map the cosmic microwave background emission at millimeter wavelengths, but also carried the Diffuse IR Background Experiment (DIRBE) to map the all-sky infrared emission over a wide wavelength range. The European Space Agency (ESA) launched the Infrared Space Observatory (ISO) in late 1995, which covered wavelengths between 2.5 and 240 microns. It covered a wider spectrum than IRAS, was thousands of times more sensitive, and had a remarkable improvement in resolution. Rather than surveying the sky, ISO carried out targeted observing programs—a measure of how well-known the infrared sky was after IRAS. The ISO mission ended in 1998.

Two other small infrared observatories were launched in the 1990s.[37] The InfraRed Telescope in Space (IRTS) was a 15 cm infrared telescope placed into orbit in early 1995, operating for about 28 days. This observatory was sponsored by the Institute for Space and Astronautical Science in Japan and eventually was retrieved during a Space Shuttle mission.

The second was the Mid-course Space Experiment (MSX), a US military satellite (launched April 1996) that included an infrared science instrument called SPIRIT-III, whose purpose was to provide higher-resolution infrared maps of crowded parts of the sky, such as the Galactic Plane and the Orion region. These maps were useful for military purposes,[38] but were also released to the scientific community.

In early 1999, the Wide-field InfraRed Explorer (WIRE) mission failed—its telescope cover was ejected prematurely, causing sunlight to vaporize the on-board solid hydrogen coolant. The Spitzer Space Telescope (SST) was another infrared telescope of the "observatory" type, carrying out targeted observing programs,

---

[37] Neither of these telescopes are currently functioning.

[38] The mission was sponsored by the US Ballistic Missile Defense Organization.

Figure 1.13. Artist's rendering of the SST. Image credit: NASA/JPL-Caltech.

proposed by different scientific teams (Figure 1.13). It was launched in 2003, and ran out of liquid helium in 2009. Even after running out of cryogen and warming up, it was still able to observe at its shortest wavelengths, and continued to do so until the mission was finished in 2020.

The Wide-field Infrared Survey Explorer (WISE) used the same solid hydrogen coolant as WIRE, but avoided its fate, generating the highest-sensitivity survey maps of the sky to date at relatively short IR wavelengths (3.3–23 microns) over 10 months in 2010. The warmed-up telescope is still operating as NEOWISE, concentrating on finding solar system objects. The Japanese Akari satellite similarly carried out a high-sensitivity sky survey at longer wavelengths in 2006–2007.

In the quarter century of orbital infrared telescope development from IRAS, telescope aperture remained very stable, ranging from 0.3 m to 0.85 m, while new detector technology delivered huge sensitivity gains. A major step forward in aperture was taken by the ESA's Herschel Space Observatory, active from 2009 to 2013, carrying a 3.5 m mirror and instruments sensitive to the far-infrared and submillimetre-wave bands (55–672 microns).

The next great step in space infrared telescopes is to be the James Webb Space Telescope (JWST), which will use a 6.5 m aperture deployable mirror to observe near- to mid-IR wavelengths. The telescope is cooled by shading it from sunlight, and launch is planned for 2021, with an "observatory" operating model.

The smaller Nancy Grace Roman[39] Space Telescope, planned for launch in 2025, is closer to the old survey missions. It will undertake a large scale near-IR survey of the universe to study dark energy cosmology, and will couple that instrument with an optical-wavelength coronagraph to search for extrasolar planets.

1.3.3 Radio Telescopes

The Birth of Radio Astronomy

In the early 1930s, Karl Jansky,[40] an engineer at Bell Telephone Laboratories, characterized the sources of radio interference ("static") that affected radio and telephone services. Jansky put together an array of dipole antennae and reflectors on a turntable that could receive shortwave 20.5 MHz signals; the structure had a diameter of 30 m and was 6 m high. Apart from thunderstorms, he found a cyclic "faint hiss" with a period of about 23 hr and 56 minute—one sidereal day, implying an extraterrestrial origin. Jansky determined that the source of this latter signal was from the constellation of Sagittarius—he had detected radio emission from the center of the Milky Way.

Jansky's pioneering work was greatly expanded by Gröte Reber (1911–2002), an amateur who built his own 9 m radio telescope. Reber's telescope is clearly recognizable as the forerunner of modern ones, being partially steerable with a parabolic reflector to define and focus a beam of radio reception on the sky. Reber used this telescope to map the radio sky in the 1930s and 1940s, discovering sources such as Cygnus A and Cassiopeia A, and showing that the radio emission was strong at low frequencies, a phenomenon later explained by synchrotron emission.

Single-dish Radio Telescopes

The progress of radio astronomy was greatly accelerated by the development of radio and radar technology during World War II, which brought many new talents to the field. One of these was Sir Bernard Lovell[41] who designed, built, and operated the 76 m diameter radio telescope at Jodrell Bank, near Manchester in the UK. This telescope had many similarities to Reber's radio telescope: a parabolic metal dish intercepted radio waves over a wide area and reflected them back into the prime focus of the telescope, where they could be detected, in analogy to an optical reflecting telescope. The focus housed a single detecting element, which measured the radiation averaged over the whole beam of the telescope. Maps of radio emission over the sky could be made by steering the telescope and hence the beam to point at different locations, as done by Reber.

Larger single-dish telescopes gave two advantages: the larger parabolic dish intercepted more radiation, so had higher sensitivity, and the size of the beam got smaller as the dish got larger, giving finer angular resolution over the sky. Several more large telescopes were therefore built: the 64 m Parkes radio telescope in New

---

[39] Astronomer and NASA's first female executive (1925–2018).

[40] Karl Guthe Jansky (1905–1950), American physicist and radio engineer. The unit of radio flux is called the Jansky (Jy; 10^{-26} W m^2 Hz^{-1}) in recognition of his work.

[41] Bernard Lovell (1913–2012), British physicist and radio-astronomer.

Figure 1.14. Australian Commonwealth Scientific and Industrial Research Organisation (CSIRO)'s Parkes radio telescopes. The indigenous name is Murriyang, which represents the "Skyworld" where a prominent creator spirit of the Wiradjuri Dreaming, Biyaami (Baiame), lives.

South Wales (Figure 1.14), Australia, the 100 m Effelsberg telescope in Germany, and the 90 m Green Bank Telescope, replaced by a 100 m instrument in 2001.

The Arecibo observatory in Puerto Rico, USA (1963–2020), adopts a different approach, suspending a 300 m dish in a karst depression, which then has limited beam steerability. The same design was used by the 500 m FAST telescope in southwest China (2020).

Radio Interferometry

Because the very long wavelengths of radio waves demand very large telescopes, telescope construction can quickly become impractical. However, by using the techniques of interferometry, a very large telescope can effectively be synthesized from multiple smaller telescopes. The technique of *Earth rotation aperture synthesis*, developed by Martin Ryle[42] can achieve the angular resolution of a telescope with diameter equal to the longest separation (or *baseline*) between the smaller telescopes, but it sacrifices the collecting area and full imaging performance of such a large telescope.

After the completion by Ryle's group of the Cambridge One Mile Telescope in 1964, the Molonglo Observatory Synthesis Telescope (MOST) was built in 1965 in Australia, and the Westerbork Synthesis Radio Telescope (WSRT) in 1970 in the Netherlands. The US Very Large Array (VLA) was completed in 1980, and the UK MERLIN network around the same time. The VLA and The Multi-Element Radio

---

[42] Sir Martin Ryle (1918–1984), British physicist and radio-astronomer.

Linked Interferometer Network (MERLIN) represent very different developments of the radio interferometer from the same basic blueprint. The VLA deploys 27 identical telescopes along railway tracks spanning about 30 km, whereas MERLIN had no more than seven heterogeneous telescopes spread out over more than 100 km. The Australia Telescope Compact Array (1980) comprises six telescopes over baselines of up to 6 km.

Very Long Baseline Interferometry (VLBI): The approach of a sparsely-filled array delivering high resolution is taken to an extreme by the field of VLBI. VLBI arrays consist of a few telescopes spread over continental or intercontinental scales. These baselines are so long that the telescope signals cannot be combined (or *correlated*) in real time, but are instead recorded for later correlation off-line. The US Very Long Baseline Array (VLBA) is spread over the US mainland and islands, while the Australian LBA covers most of that continent. Continent-scale arrays are also found in Europe (The European VLBI Network (EVN)) and East Asia. Smaller arrays, such as the Korean and Japanese VLBI Network (JVN), are more constrained in their space. In general, VLBI techniques are capable of resolving radio emission on angular scales of milli-arcseconds, orders of magnitude finer than the resolution of optical ground-based telescopes.

The Move to Higher Frequencies

Observing at higher frequencies requires technological development, as receivers need to be more precisely built, telescope reflector surfaces need to be more accurate, and electronic systems need to work at the higher frequency. These developments became available over the second half of the 20th century, and a corresponding move toward higher frequencies can be seen in radio astronomy. At frequencies higher than a few GHz, synchrotron emission drops out of view, and the radio sky is dominated by the thermal bremsstrahlung radiation (see Section 6.1.1) from plasmas, and then blackbody radiation from interstellar dust grains at higher frequencies still. Superimposed on these broadband emission mechanisms are the spectral lines caused by quantum transitions of interstellar molecules and atoms. These transitions can be used to estimate the physical conditions of interstellar gas, and some of them (e.g., carbon monoxide and atomic carbon transitions) are important to the overall physics of galaxies.

While some molecular line transitions had been detected at fairly low frequencies in the 1960s, in 1970 the rotational ground-state transition of carbon monoxide (CO $J = 1 - 0$) was discovered at frequency at a frequency of 115 GHz, or 2.6 mm wavelength. This marks one of the beginnings of the field of *millimeter-wave* radio astronomy (or mm-wave), which in turn gave rise to *submillimetre-wave* radio astronomy (or sub-mm) at even higher frequencies.

Many early mm-wave telescopes were small by radio telescope standards—a few meters in aperture. The Texas Millimeter Wave Observatory used a 4.9 m antenna from 1971. From 1974, the 1.2 m Columbia Millimeter Wave Telescope surveyed the CO $J = 1 - 0$ emission of the Milky Way from a New York City rooftop; a twin instrument was used in Chile from 1982. Even in the 1990 s, the Japanese NANTEN 4 m telescope was used for sky surveys in Chile.

The National Radio Astronomy Observatory (NRAO) 12 m telescope in Arizona was used extensively for mm-wave work, and many more 10–15 m class telescopes followed it, in Massachusetts (FCRAO), Sweden (Onsala), Chile (SEST), and other locations. A few larger-aperture telescopes were also used: the 30 m Pico Veleta telescope (Spain); the 45 m Nobeyama Radio telescope; even the 100 m Effelsberg telescope could be used up to 95 GHz. In the 21st century, the 50 m Large Millimeter Telescope (Mexico) and 100 m Green Bank Telescope (USA) were completed.

Aperture synthesis was also applied to mm-wave radio astronomy, with interferometric arrays built in California (the Owens Valley Radio Observatory and the Berkeley–Illinois–Maryland Array) and in France (the Plateau de Bure Interferometer). The Australia Telescope Compact Array (ATCA) interferometer (Australia, see above) was also fitted with receivers to allow it to operate up to 110 GHz (Figure 1.15).

Submillimetre-wave Radio Astronomy and ALMA: In the 1980s, the high-frequency limit was pushed past about 300 GHz, and into the sub-mm spectrum. To observe such high frequencies, telescopes needed to be purpose-built at high, dry sites. Early submillimetre-wave telescopes were the 3 m KOSMA telescope (Switzerland), the 10 m Caltech Submillimeter Observatory and the 15 m James Clerk Maxwell Telescope (both in Hawaii). The 10 m Heinrich Hertz Telescope (Arizona) and 12 m APEX telescope (Chile) were both built in the 1990s.

Figure 1.15. The Australia Telescope Compact Array (ATCA) at the Paul Wild Observatory near Narrabri, Australia.

In the 21st century, the first spectroscopy at terahertz (>1000 GHz) frequencies was undertaken, mainly from the highest sites in Chile, and from Antarctic plateau observatories. The 1.7 m AST/RO and 10 m SPST telescopes at the US Amundsen–Scott South Pole Station, and the HEAT telescope at the remote Ridge A plateau site, were used for submillimetre-wave and terahertz observations. Terahertz observations have also been undertaken by long-duration balloon flight.

The Atacama Large Millimeter Array (ALMA) is an interferometer comprising 54 12 m telescopes and 12 7 m telescopes on the 5000 m high Chajnantor Plateau in the Atacama desert of Chile. It can observe essentially the whole spectrum from 100 to 1000 GHz.

The Square Kilometre Array

The Square Kilometre Array (SKA) radio telescope is currently under construction. It aims to have high angular resolution by using aperture synthesis, but also to have a large collecting area (of order a square kilometer), and many baselines, by constructing the array of many small telescopes. The array is planned to consist of an array of traditional parabolic reflector telescopes in South Africa to observe frequencies of a few GHz, and an array of dipole antennae in Australia to observe lower frequencies. The SKA is such a large project that it requires several "path-finder" projects to develop and demonstrate new technologies. These pathfinders are themselves substantial observatories, such as LOFAR in the Netherlands, MeerKAT in South Africa, and ASKAP and the MWA in Australia. In particular, LOFAR and MWA have demonstrated new low-frequency observing technologies, while ASKAP has demonstrated the use of Phased Array Feeds, which allow radio signals to be detected in multiple "pixels" at once, greatly improving the area coverage of the telescope.

The Cosmic Microwave Background

Since its discovery in 1965, observations of the Cosmic Microwave Background (CMB)—the redshifted radiation from the opaque primordial universe—have generally used purpose-built telescopes designed to make specific measurements of the radiation. These telescopes generally work at wavelengths of millimeters to about a centimeter, where the CMB spectrum peaks. While centimeter-wave CMB telescopes have been sited in many places, mm-wave ones generally are sited at the same kinds of locations as submm-wave telescopes: the Atacama Desert and the Antarctic Plateau, due to their very tight requirements on atmospheric noise.

Radio Astronomy from Space

Space-based radio astronomy projects are quite rare, since the atmosphere is mainly transparent to radio waves, and radio telescopes are generally rather large. However, one obvious use for space-based radio astronomy is to achieve long baselines for interferometry—space VLBI. In the 1990s, the Japanese VSOP project demonstrated VLBI between ground and space antennae. In 2011, the Russian Radio-Astron satellite was launched into an elliptical orbit that could achieve baselines of up to 350,000 km, but the spacecraft became unresponsive in 2019.

Space-based telescopes have made more impact in the mm-wave and sub-mm fields, where the Earth's atmosphere is a serious impediment to observations. Space-based sub-mm spectroscopy was carried out by the dedicated SWAS and ODIN satellites, and was part of the European Herschel Space Observatory's mission.

Although the Earth's atmosphere is fairly transparent in the millimeter-wave spectrum where the CMB peaks, it still introduces enough noise to be a limiting factor on the mapping and analysis of the CMB, so space-based CMB telescopes have made great strides in our understanding of cosmology. The first of these was Cosmic Background Explorer (COBE; 1989), followed by the US Wilkinson Microwave Anisotropy Probe (WMAP; 2001) and the European Planck (2009) missions.

1.3.4 X-ray Telescopes

The Earth's atmosphere is fully opaque to X-rays, so X-ray astronomy must be done from the edge of space (using balloons) or from space with rockets and satellites. The development of X-ray astronomy is therefore, a succession of space missions from the dawn of the Space Age to today. Following the detection of X-rays from the Sun (with a captured V-2 rocket), the first cosmic X-ray sources were discovered in 1962—Scorpius X-1 and the cosmic X-ray background. This led to the development of large-area imaging X-ray detectors and telescopes. X-ray imaging telescopes were invented in the 1950s by Hans Wolter,[43] using a two-mirror configuration. These mirrors were designed so that X-rays are reflected only at very shallow angles (*grazing incidence optics*), and the combination of mirrors gives acceptable image quality over a wide field.

In the early 1970s, the first X-ray telescope was flown: the Apollo Telescope (mounted on the US Skylab space station). The Uhuru satellite (originally called SAS-1) used X-ray imaging detectors to perform the first X-ray sky survey, finding 339 sources. The German Balloon-HEXE instrument discovered a cyclotron resonance line in the spectrum of the X-ray binary pulsar Hercules X-1, which yielded a measurement of the magnetic field strength of the neutron star of 500 million Tesla.[44]

In the late 1970s and 1980s, larger satellites carried the new X-ray telescopes into orbit for large-scale programs. HEAO-2 (renamed the Einstein Observatory) and the European EXOSAT both carried Wolter telescopes. The 1990s saw the German–UK–US Röntgensatellite (ROSAT) mission carry out the deepest sky survey to date, while the Japanese–US Advanced Satellite for Cosmology and Astrophysics (ASCA) and Italian–Dutch BeppoSAX[45] satellites carried out more specialized tasks—in particular, BeppoSAX was able to identify the X-ray counterparts to gamma-ray bursts.

---

[43] Hans Wolter (1911–1978), German physicist.

[44] Unit named after Serbian scientist Nikola Tesla (1856–1943). One Tesla equals one Weber per square meter, corresponding to 10^4 Gauss.

[45] Named after Italian physicist Giuseppe "Beppo" Occhialini (1907–1993).

Figure 1.16. The X-ray Spectrum–Röntgen–Gamma mission to survey the whole sky at 0.2 to 30 keV energies. There are two instruments on-board of Spektr-RG: One is eROSITA—an X-ray survey telescope designed and built by Germany's Max-Planck-Institut für extraterrestrische Physik (MPE) in Garching, which will operate at 0.2–10 keV. The second one is ART-XC—a Russian-built telescope that can detect "harder" X-rays at 6–30 keV energy range. Image credit: Peter Friedrich, MPE.

X-ray astronomy at the start of the 21st century was dominated by two large orbiting observatories—the European XMM-Newton and the US Chandra[46] satellites. Both of these satellites carried out varied observing programs proposed by astronomers. Two decades into the century, a new sky survey will be carried out to supersede ROSAT's: The Russian Spectrum–Röntgen–Gamma satellite carries the German eROSITA X-ray telescope, and launched successfully in July 2019 (Figure 1.16). If eROSITA can be seen as a successor mission to ROSAT, the successor to XMM-Newton and Chandra is ATHENA, an observatory that is to be an order of magnitude more sensitive than its predecessors, and is currently planned for launch in 2031.

1.3.5 Cosmic Rays, Extensive Air Showers, and Čerenkov Radiation Telescopes

Before moving up the electromagnetic ladder to gamma-rays, we take a moment to discuss cosmic rays and the role they played in the understanding and development of high energy detection techniques.

Cosmic rays are a mix of high energy particles, mostly (90%) protons having energies up to 10^{21} eV (see Sections 3.2.1 and 7.6). Traveling at near the speed of

---

[46] Named after Indian astrophysicist and Nobel Laureate Subrahmanyan Chandrasekhar (1910–1984).

COSMIC RAYS MAY FORECAST WEATHER

Dr. R. A. Millikan at work with his latest electroscope, with which he is studying the cosmic rays. He believes these mysterious rays may be used in making reliable forecasts of the weather.

Cosmic rays may help to prophesy the weather. This first practical use for the mysterious radiations from outer space was recently announced by Dr. R. A. Millikan, Calif. Institute of Technology physicist.

The "cosmic rays" are more penetrating than radium or X-rays, but it is not known whether they affect human beings.

Dr. Millikan, who discovered the source of the rays (P. S. M., July, '28, p. 13), has measured their strength with his new electroscope, and is able to determine high-altitude atmospheric conditions.

Page 32 Popular Science Monthly March 1931

Figure 1.17. Interesting suggestion by Robert A. Millikan (1868–1953) that cosmic rays may forecast weather. Image credit: Popular Science Monthly (March 1931).

light, they create extensive air showers when they encounter the atmosphere. The discovery of these particles is generally attributed to Hess,[47] who flew charged electroscopes in balloons high into the atmosphere; he found the rate of electroscope-discharge increased significantly above 2000 m altitude. At the time, interest about the importance of cosmic rays came from a number of scientists (see for example Figure 1.17).

Hadronic showers are created by these charged high energy cosmic ray particles impacting the atmosphere. These are composed of pions and kaons that decay into muons and muon-neutrinos. Showers are also created by high energy gamma-rays (termed gamma showers); these have a different signature such that they can be distinguished from hadronic showers by modeling (see Völk & Bernlöhr 2009 for a discussion of narrow, more directed gamma-ray showers versus broader, more irregular hadronic showers).

By the time these showers reach the ground, they consist of muons and electron/positron pairs in a ratio of 3:1 traveling at $0.998\,c$ with energies greater than 4 GeV. When these charged particles travel through a medium faster than the speed of light through that same medium, a bluish light is emitted known as Čerenkov radiation[48] (Section 6.1.3).

---

[47] Victor Hess (1883–1964), German physicist; Nobel laureate in Physics 1936.
[48] discovered by Russian physicist Pavel Čerenkov (1904–1990) in 1934.

The exact form of these showers, as revealed by their Čerenkov radiation, allows the energy and direction of the original cosmic ray to be estimated. In this way, the material in which the Čerenkov radiation is emitted becomes a telescope for very high energy particles. These telescopes are found on the ground, underground, and deep in oceans, as we will discuss in the next section.

Cosmic rays are found from all directions of the sky and may be associated with high energy events such as supernova (SN) explosions, massive stellar core mergers and active galactic nuclei (AGN). They may also be associated with normal stars such as our Sun. Since cosmic rays are mostly protons and other charged particles, they are directionally affected by the electromagnetic fields in both the Galaxy and our solar system. This makes true astronomy with cosmic rays extremely difficult except for the highest energy particles ($\geq$ EeV range or $\geq 10^{18}$ eV), where it is possible the determine their direction back to their source.

1.3.6 Gamma-ray Telescopes

Gamma-rays at MeV–GeV energies have been typically observed with space-based instruments, but at higher energies those instruments are ineffective. In late 1980s, with the advancement of the Imaging Atmospheric Cherenkov Telescopes (IACTs), ground-based observation of TeV gamma-rays came into reality and, since the first source detected at TeV energies in 1989, the number of gamma-ray sources has rapidly grown.

The discovery of the famous remnant of the Supernova 1054 AD explosion—the Crab Nebula (Messier 1 or M 1)—as the first source of TeV gamma-rays in 1989 has marked the start of ground-based observational gamma-ray astronomy in the very high energy range of 100 GeV–10 TeV. Led by Trevor Weekes (1940–2014), after twenty years of trial and error, success and misfortune, step-by-step improvements in both the technique and understanding of gamma shower discrimination methods, using the 10 m diameter telescope on Mount Hopkins in Arizona, finally succeeded in measuring a 9σ gamma-ray from the direction of the Crab Nebula.

Gamma-ray astronomy (Leahy et al. 2018) is basically the astrophysics of nuclear particles and elementary particles from sources in extreme gravitational, electromagnetic and thermal conditions. Space observations are required since energies below 1 TeV cannot be properly detected on ground-based instruments. The photoelectric effect is dominant <100 keV, Compton[49] scattering between 100 keV and 10 MeV, and electron–positron pair production at energies above 10 MeV.

The strongest gamma-ray sources include the Sun and some gamma-ray bursts. Directionality can be achieved by shielding and masking (lower energies), or at higher energies, by using directional properties of the Compton effect or pair production. Although the angular resolutions of these telescopes are only a fraction of a degree, this disadvantage is offset by a non-crowded gamma sky and source identification by time variations.

---

[49] Named after physicist Arthur Holly Compton (1892–1962) who won the Nobel Prize in Physics in 1927.

The first satellite to detect gamma-rays or hard X-rays was Explorer XI (1961); it detected 22 hard X-rays. In 1967, the Third Orbiting Solar Observatory (OSO-3) detected diffuse cosmic hard X-rays, solar flares, and a single flare from Sco X-1 before becoming useless in 1968. OSO-7 detected the first solar flare gamma-ray lines in 1972 and in the 1980s, 9 years worth of all kinds of solar gamma-ray phenomena were detected by the Solar Maximum Mission. Vela satellites, designed to detect man made nuclear activity, detected gamma-ray bursts in 1967, which initially caused quite a stir among the military.[50]

Improvements in technology included spark chambers to record pair production at photon energies >30 MeV and Compton telescopes which observed in the 1–10 MeV range. Attempts to observe the Galaxy for high energy gamma-rays using SAS-2 failed due to a malfunction of tape recorders. COS-B was more successful (1975–1982); it used a wire spark chamber and a total absorption counter to measure energy. It produced the first sky map of the Milky Way continuum emission. This emission is created by interactions of cosmic rays with interstellar matter, Supernova Remnants (SNRs), PulsarWind Nebulae (PWN), and pulsars (and possibly other yet unidentified objects).

The Compton Observatory was launched by a space shuttle mission in 1991 and operated for 9 years. The satellite included the wire spark chamber Energetic Gamma-ray Experiment Telescope (EGRET) for energies >30 MeV, a large Cesium Iodide crystal spectrometer, a Compton Telescope (COMPTEL) for 1–30 MeV, the gamma-ray Burst and Transient Source Experiment (BATSE), and the Oriented Scintillation-Spectrometer Experiment (OSSE). The SIGMA mission (1989) augmented Compton observations by studying the gamma-ray Galactic Center.

This was followed by ESA's INTErnational Gamma-Ray Astrophysics Laboratory (INTEGRAL) launched in 2002. INTEGRAL provided a better understanding of gamma-ray lines produced from radioactive element decay, outbursts from X-ray binaries, and gamma-ray counterparts to gravitational wave binary neutron star mergers.

Other achievements of gamma-ray astronomy include the study of diffuse emission (interactions of cosmic rays with matter and matter–antimatter annihilation), nuclear lines (solar gamma-rays), localized sources (pulsars, AGN), gamma-ray burst sources (compact relativistic sources) and other as yet unidentified sources.

Since 2003, as the new generation experiments (HESS, MAGIC, CANGAROO, and VERITAS) started to observe the gamma-ray sky, the number of VHE sources started to rapidly increase. A new class of sources was detected at GeV–TeV energies from both galactic sources (e.g., Galactic Center, PWN, SNRs, pulsars and various binary systems) and extragalactic sources (such as blazars, radio galaxies and star-forming galaxies) as well as about a dozen unknown new TeV sources.

Using this technique we discovered over several hundred of gamma-rays of very different types—both, galactic and extragalactic origin. There is no doubt that this is a fast evolving branch of astrophysics which is rapidly improving our understanding

---

[50] In the mid-1990 s, the extragalactic origin of gamma-ray bursts was determined by BeppoSax.

Figure 1.18. CTA Telescopes in Southern Hemisphere. All three classes of the telescopes planned for the southern hemisphere at ESO's Paranal Observatory, as viewed from the center of the array. Image credit: CTAO/M.-A. Besel/IAC (G. P. Diaz)/ESO.

of the most violent and energetic sources and processes in the sky. The brief history of ground-based very high energy gamma-ray astrophysics with atmospheric air Čerenkov telescopes is comprehensively described in Mirzoyan (2014). Also, more information about Čerenkov telescopes can be found in Section 7.8. The reader is also referred to a fascinating history of the evolution of TeV Astronomy by Fegan (2019).

The future of this field is the Cherenkov Telescope Array (CTA), which will be the major global observatory for very high energy (VHE) gamma-ray astronomy over the next decade and beyond (Figure 1.18). Covering photon energies from 20 GeV to 300 TeV, the CTA will improve on all aspects of atmospheric imaging techniques using Čerenkov (IACT) instruments, allowing detailed imaging of a large number of gamma-ray sources, and being a powerful instrument for time-domain astrophysics. The CTA will be built in two places: the Canary Islands (for the surveys of the northern hemisphere) and the Atacama Desert in Chile (southern hemisphere array).

1.3.7 Neutrino Astronomy: The Early Days

The history of high-energy astrophysical neutrino telescopes begins with the Deep Underwater Muon And Neutrino Detector (DUMAND) Project in 1976. By then, observations of neutrinos with MeV energies from fusion processes in the Sun were underway, which would eventually lead to the discovery of non-zero neutrino masses and neutrino oscillations. It had also been realized, however, that the interactions of cosmic rays with gas and radiation fields in the universe would produce a flux of neutrinos with energies up to tens of EeV (1 EeV = 10^{18} eV; Appendix A). Upon interaction, these neutrinos would generate charged secondary particles that would then emit optical–UV Čerenkov light (the electromagnetic equivalent of a sonic boom), which could be used to reconstruct the energy and direction of the neutrino primary.

High-energy astrophysical neutrinos, however, would be much rarer than solar neutrinos and require a target volume of order 1 km^3 to study these neutrinos in the TeV–PeV range. This is the energy range expected from neutrino secondaries which are the result of cosmic ray interactions in our galaxy. Constructing such a large detector however was completely out of the question. The solution was to find a large, dark, optically transparent volume of natural material in which to install light sensors to search for secondary particles from neutrino interactions.

DUMAND's target volume was located in deep ocean water at a seabed depth of 5 km off the Hawaiian coast. Its aim was to instrument 1 km^3 of seawater using optical detection modules. The project, however, was overcome by technical difficulties and was officially canceled in 1995.

Construction of detectors at Lake Baikal in Russia and South Pole Station in Antarctica began in the early 1990s. The Baikal Deep Underwater Neutrino Telescope (BDUNT) was completed in 1998 with 192 optical sensors, while the Antarctic Muon And Neutrino Detector Array (AMANDA) with 667 optical sensors was completed in 2000. While these detectors were too small to identify astrophysical neutrinos against the background of neutrinos from cosmic ray interactions in the atmosphere, their successful operation allowed the development of detector technology and analysis techniques which would be vital for the next generation.

Meanwhile, in the Mediterranean, three neutrino telescope projects were taking shape. These were ANTARES off the southern coast of France; NEMO, near the South-East tip of Sicily; and NESTOR, near Pylos, Greece. ANTARES was the most advanced of these projects, being completed in 2008 with 885 optical sensors with a size similar to AMANDA. It is still operating, but is due to be decommissioned in the near future.

The first, and currently only, neutrino telescope to reach the milestone of 1 km^3 is the IceCube Neutrino Observatory. Begun in 2005 in a volume of ice surrounding the AMANDA detector and completed in 2020, IceCube consists of 5160 optical modules on 86 lines at depths between 1.4 and 2.4 km. It has since been implemented with an infill detector, Deep Core, to study neutrinos down to tens of GeV of energy and the IceTop surface array to detect incoming cosmic rays. IceCube's two major breakthroughs have been the discovery of the first astrophysical neutrinos in 2013 and the 2018 announcement of neutrinos from the blazar TXS 0506+056, bringing high-energy astrophysical neutrino detection into the realms of astronomy.

The future of neutrino astronomy is bright. The Gigaton Volume Detector (GVD) is under construction in Lake Baikal, while Phase 1 of the Cubic KiloMetre (km3) Neutrino Telescope (KM3NeT; Figure 1.19)—which has grown out of the ANTARES, NEMO, and NESTOR collaborations—is under construction at two

Figure 1.19. The KM3NeT project—observing neutrinos from the universe. View from inside a detector building block of detection units. The yellow buoy of a nearby detection unit is clearly visible. Image credit: Edward Berbee/Nikhef.

sites in the Mediterranean. Both have a target volume of ~ 1 km^3, with completion scheduled for the next decade. IceCube is also planning a next-generation detector, with a total volume of 10 km^3. Together, these instruments promise to detect further sources of astrophysical neutrinos and open a new window on the universe.

1.3.8 Brief History of Gravitational Wave Astronomy

Perhaps the earliest conception of gravitational waves is to be found in Maxwell's musings in 1865 (Maxwell 1865) about the possibility of gravity being a field similar to electromagnetism since they each were described by the inverse square law. Heaviside[51] (1893) in 1893 examined this idea as a "Gravitational analogy": recasting electromagnetic theory in terms of mass, inertia, and gravitational force as described by Newton's law. He also showed that gravitational force propagates at a single finite speed even though Newton did not accept force at a distance without an intervening medium (Newton 1758).

With the publication of the special theory of relativity (1905), Albert Einstein (1905) completed the merger between electricity and magnetism which was started by Maxwell. The propagation of gravity could not be faster than the speed of light and was equal, in fact, to the speed of light. Nearly contemporaneous with Einstein's[52] paper was one by Jules Henri Poincaré (1905),[53] which presented very

---

[51] Oliver Heaviside (1850–1925) was an English self-taught electrical engineer, mathematician and physicist.
[52] Albert Einstein (1879–1955) was a German-born theoretical physicist.
[53] Jules Henri Poincaré (1854–1912) was a French mathematician, theoretical physicist, engineer, and philosopher of science.

similar results but lacked important features of Einstein's theory. Poincaré showed that, since gravitational force propagates at the speed of light, there will be a time lag between a change in gravity and its effect. This delay due to the finite propagation velocity of gravity he considered as the possible source of gravitational waves (Leverenz & Filipović 2021).

Founded upon Einstein's theory of relativity, the youngest branch of multimessenger astronomy might revolutionize our understanding of the universe. It is widely accepted that gravitational waves were predicted by Einstein in 1916 as a specific consequence of his general theory of relativity. However, it quickly became a common feature of all modern theories of gravity that obey special relativity (Schutz 1984).

A long debate started in 1916 around whether the waves were actually physical or were artefacts of coordinate freedom in general relativity. This was not fully resolved until the 1950s. Until his death in 1955, Einstein's skepticism regarding the reality of gravitational radiation impeded research into gravitational waves, particularly the question of whether or not they could be detected.

The Early Study of Gravity

The field of gravity research was re-invigorated in the 1950s by Agnew Bahnson.[54] He sponsored an essay contest through the Gravity Research Foundation (GRF) on the possibility of insulators, reflectors, or absorbers of gravity. In 1953, Bryce DeWitt entered an essay contest conducted by the GRF with a submission that dismissed the entire anti-gravity concept. He won the contest and a prize of $1000 (US). Dewitt, supported by several senior physicists including Oppenheimer, Dyson, Teller, Feynman, and Wheeler, then helped to establish something akin to an "Institute for Advanced Study" for gravity research with real scientific goals, distancing the new institute from the GRF. As a group, they were very concerned that gravity research had been neglected for many years. The details are fascinating and presented in the report from the 1957 Chapel Hill Conference Report (DeWitt & Rickless 2011). Any mention of anti-gravity was scrupulously avoided to maintain the legitimacy of the organization and in order not to discourage sponsors and scientists.

The 1957 Chapel Hill conference had 40 speakers from 11 countries and met for six days. A critical discussion in that conference concerned the effect of a gravitational pulse on a particle and whether or not the wave would transmit energy to the particle. Feynman reasoned, through a thought experiment now referred to as the "sticky bead argument," that heating would occur and energy would be deposited in the system, giving rise to the expectation that gravitational waves could be detected, at least theoretically.

Detecting Gravitational Waves

Among the attendees of the 1957 Chapel Hill meeting was Joseph Weber (1919–2000), who was an engineer at the University of Maryland. He published an

---

[54] Agnew Bahnson (1915–1964) was a wealthy North Carolinian industrialist who had a passion for gravitational physics, especially anti-gravity.

approach to gravitational wave detection using a mechanical system with piezo-electric crystals in 1960 (Weber 1960). Results of an experiment with two large, resonant, metal cylinder "bar" gravitational wave "antennae" were published in 1969 (Weber 1969). The experiment detected coincidence in two detectors with ~1000 km separation. These positive results have never been replicated. In 1972, Kafka (1972) suggested that the mass loss to the Galactic center would need to be ~3×10^3 M$_\odot$ yr^{-1} in order to be detected by this bar experiment whereas the expected mass loss is a few hundred M$_\odot$ yr^{-1}.

Interferometric detection of gravitational waves was considered starting in the 1960s (Forward 1978). The basic design was a folded Michelson interferometer with environmental isolation including vibration isolation from the ground and with all optical components mounted in a vacuum. During the 1960s through the 1980s, several gravitational waves experiments were conducted which included the construction of instruments with sizes of 3 m to 30 m (Cervantes-Cota et al. 2016) and with unrealized proposals for up to 3 km designs. Much was learned and techniques were developed as progress was made toward a viable instrument including the noise suppression methods later used in the Laser Interferometer Gravitational Wave Observatory (LIGO) project (De Sabbata & Weber 1977).

In the late 1980s we got the first indirect observational evidence for the existence of gravitational waves. This came from monitoring of the PSR B1913+16 (also known as Hulse–Taylor or PSR J1915+1606) binary pulsar (originally discovered in 1974). The pulsar's orbit was found to evolve exactly as would be expected for gravitational wave emission (Hulse & Taylor 1975). In 1993, the Nobel Prize in Physics was awarded to Russell Alan Hulse (1950–) and Joseph Hooton Taylor Jr (1941–) for this discovery.

In the 1990s major advances were made with a 600 m interferometer designed in Germany. The GEO 600 interferometer construction began in 1995. This instrument has been in operation since 2002 with an extended LIGO-GEO science run in 2005.

The LIGO organization was created by a 1984 agreement between Caltech and MIT LIGO (Thorne & Weiss 2016).[55] The plan was to build and operate a pair of "initial interferometers" based on existing technology followed by "advanced interferometers" utilizing the newest technology developed for the experiment. In 1997 the LIGO Scientific Collaboration (LSC) was established. It is responsible for operations, advanced interferometer research and development, and the expansion of technical and scientific cooperation beyond Caltech and MIT.

By 2016 approximately 1000 scientists at 75 institutions in 15 nations were participating. The first gravitational wave search using the initial interferometers was performed in 2002–2005. In 2007 coordination was established with the European interferometric gravitational waves experiment Virgo, named for the Virgo galactic cluster.

On 2015 September 4, during the commissioning and testing of the advanced LIGO detectors, gravitational waves were detected directly for the first time.

---

[55] https://www.ligo.caltech.edu/system/media_files/binaries/313/original/

Figure 1.20. Left: the Einstein telescope (http://www.et-gw.eu/index.php) and right: LISA, space-based interferometer (https://lisa.nasa.gov/). Images credit: left: NASA/C. Henze, right: NASA.

In recognition of this monumental triumph, the project leaders for the LIGO experiment were awarded the 2017 Nobel Prize in Physics.

Since then, further improvements have been made in sensitivity. Additionally, the number of interferometers in use has increased. As more interferometric detectors become available, sensitivity and source position accuracy will be greatly improved. Wide geographic distribution of the interferometers enables improved localization of the sources detected. In 2020 two detectors are operating in the USA: Hanford in Washington state and Livingstone in Louisiana state. Three interferometers operating on other continents are the Virgo interferometer near Pisa, Italy; GEO 600 near Hannover, Germany; and KAGRA located underground in Japan. Planned for operation as early as 2025 is LIGO-India. Future interferometers include the Einstein telescope and LISA (Figure 1.20).

The number of gravitational wave detections continues to grow. The web site in LIGO Scientific Collaboration (2021)[56] can be consulted for the current list.

1.4 Locations: From City to Space

As new technologies have been invented and adapted to increase our view of the universe, so the location of the instrumentation has become crucially important. The telescope environment defines effects on the observations that can become limiting factors.

1.4.1 What Gets in the Way?

If the signal is the information coming to our instrument from an astrophysical object, it can be degraded in two fundamental ways: By extraneous signal being added to the astrophysical one, and by degradation of the original signal.

The extraneous signals that can be added to the signal of interest can be human-generated, such as light pollution (e.g., from cities) for optical telescopes, or radio

---

[56] https://www.ligo.org/detections.php

interference (RFI, e.g., from communications and other technologies) for radio telescopes. Additional signal can also be added naturally—the atmosphere emits light at specific frequencies (aurora and *air-glow*; see Section 5.1.2) and broadband infrared radiation as blackbody emission from the air itself. Even gravitational-wave detectors have to contend with natural and man-made signals in the form of ground movements.

The original astrophysical signal can be degraded simply by absorption by the atmosphere. The atmosphere is never fully transparent, and at some wavelengths it becomes opaque. Therefore, some fraction of the incoming signal will be removed by the atmosphere, depending on the *opacity* (Section 5.3.1) of the atmosphere above the instrument. In principle, this can be corrected, but the absorption is generally not completely smooth, so that any estimate of the atmospheric absorption will include some errors. The overall effect is to add a source of noise. In addition to the challenges of estimating the magnitude of the astrophysical signal, environmental factors can degrade the localization of the signal, i.e., the image quality. This is most noticeable in the optical regime, where turbulent motions in the atmosphere move around air packets with slightly different refractive indices. The light therefore has its direction changed in an unpredictable way, and the overall effect is one of blurring—astronomical *seeing* (Section 5.2.4). A related problem is that of *scintillation*, in which the detected source seems to wander in position and vary in brightness, due to higher-altitude turbulence; scintillation is observed at radio wavelengths as well as optical (Section 5.2.4).

The easiest way to remove or alleviate environmental factors is simply to choose a better location.

Mountains
Many atmospheric effects (such as absorption and emission) scale with the amount of air between observer and source, and thus going to high altitude improves them. Some atmospheric effects—such as scintillation, seeing, and air-glow—are concentrated to particular atmospheric layers and particular physical conditions—such as the degree of turbulence—so location selection is harder than just going uphill. Greater understanding of the atmospheric physics underlying seeing has led to better and more detailed testing, and has also reduced the number of world-class mountain locations to three: the South American Andes, Hawaii, and the Canary Islands.

Polar Locations
The Antarctic Plateau lies at high altitude—due to millennia of snow accumulation, the top of the plateau is around 4000 m above sea level. But, in comparison to a mountain site of comparable altitude, it is very flat, and also much colder. The absolute humidity (measured as the amount of *precipitable water vapor* or pwv) is very low, although the relative humidity is high. These considerations make polar locations very good for some instruments: the cold air has less thermal emission, and the low water vapor reduces absorption. Although there is plenty of atmospheric turbulence, it is confined to a shallow boundary layer, leading to reduced opacity noise, and to the prospect of excellent natural seeing above that layer.

Aircraft and Balloons
Remaining in the atmosphere, aircraft- and balloon-borne telescopes obtain reduced atmospheric effects, but the effects are not eliminated. With atmospheric effects still in evidence, and strong size and weight constraints, aircraft and balloon-borne astronomy is something of a compromise solution. It has mainly found its niche in the infrared and submillimetre-wave spectrum.

Near-Earth Space
Deployment of telescopes into space fully removes atmospheric effects from the astrophysical signal. While the Hubble Space Telescope,[57] operating in UV, visible and near-IR frequencies, is the most visible of such telescopes, it is hardly the only one. From short experimental missions to long-lived multiinstrument satellites, space telescopes have revolutionized our understanding of the universe. The whole electromagnetic spectrum is visible without absorption, and atmospheric seeing is removed. The only environmental factors remaining are due to electromagnetic radiation from the Earth–Moon system, including thermal emission in the infrared, bright reflected visible light, and radio interference.

Deep Space
These final environmental effects can be largely removed by moving to orbits outside the Earth–Moon system. In particular, orbits around stable Lagrange points can be used to keep a space observatory in a well-controlled position, well away from the Earth. An example of a space telescope using this location is the infrared James Webb[58] Space Telescope, orbiting the second Sun–Earth Lagrange point (L2)—by being further away from Earth, it receives less infrared radiation from Earth, allowing it to cool down further, which in turn reduces the noise on the astrophysical signal. A variant of this strategy was used by the Spitzer[59] Space Telescope that launched into an Earth-trailing solar orbit.

Additional Benefits of Space
Most of the environmental benefits of locating instruments in space arise from the lack of factors that degrade performance. There are, however, additional opportunities in space that arise from its unique characteristics. An instrument can be made much larger in space. A ground-based VLBI baseline cannot be made larger than the Earth, but a space-based VLBI system can have much larger baselines. It is even possible to build a much larger gravitational-wave detector, such as the planned eLISA system, with arms much larger than the Earth—a scientifically exciting and technologically daunting project.

---

[57] Named for Edwin Hubble (1889–1953), American astronomer and discoverer of the expansion of the universe.
[58] James Webb (1906–1992), American administrator of NASA.
[59] Lyman Spitzer (1914–1997), American astrophysicist, who was the first to suggest that we should put telescopes in space in the 1940s.

Figure 1.21. One of the futurists' visions for the Moon-base. Image credit: NASA Innovative Advanced Concept (NIAC) project Contour Crafting; Robotic construction of Lunar and Martian infrastructure, University of Southern California Center for Rapid Automated Fabrication Technologies (CRAFT).

Lunar Observatories

There have been a number of proposals to build new instruments on the Moon; these generally assume that some infrastructure will be available due to Lunar exploration programs. Lunar telescopes could be built *in situ*, allowing much larger structures than those launched into space; for example a lunar radio telescope for VLBI would not only have a very long baseline to Earth, but could conceivably be built to have a large collecting area (Figure 1.21). Most Lunar telescope concepts, however, envisage siting their instruments on the Lunar far side, allowing the Moon to block out the signals from the increasingly noisy Earth. One proposal for an optical telescope, building a giant liquid-mirror telescope on the far side, hopes to be able to detect the very first generation of stars in our universe (Schauer et al. 2020).

1.4.2 Observatories and Experiments

The adoption of telescopes large enough to require their own infrastructure led to the operating model of the observatory, in which a large astronomical instrument is constructed in a good location, and then astronomers bring their projects and ancillary instruments to that location. Many projects of many different kinds can be carried out at the same observatory, even if it was originally built to investigate a specific problem (such as mapping the "canals" of Mars in the late 19th century).

In the late 20th century, as astronomy became much closer to physics, a different operating model started to become important. An experiment is a set of equipment and instrumentation that is designed for one project or measurement, and for nothing else. It may sometimes be repurposed to a different project later on, but this is not an element of the design. Some of the earliest examples of this model were early space-based observations from, e.g., suborbital rockets, in which the flight was very carefully planned. Many ground-based cosmic microwave background telescopes have been run as experiments, with the purpose of making a specific measurement of the background signal.

Many projects—particularly space telescopes—will combine elements of both models.

In general, the experimental approach allows the hardware and observing plan to be rigorously optimized for its use case. The observatory approach allows the instrument to be repurposed as new science cases emerge; many instruments are best-known for discoveries that were not in the original scientific plan. A hybrid approach allows a particular instrument to combine elements of the experiment and observatory approaches to suit its circumstances. A space telescope may have a few narrowly-defined observing techniques, but be able to execute them toward many different targets (such as Spitzer), or may act as an observatory, accepting proposals for instrument use but without any hardware upgrades (such as the Chandra X-ray observatory), or may even be upgradeable in orbit, like the HST.

1.4.3 Logistics

While new locations offer improved environmental factors, they also tend to come with logistical challenges, which translate to higher costs. These increased costs can often be tolerated. Many of the most modern instruments are so inherently expensive that only one can be built at a time, and the extra logistics costs involved in placing the instrument in an optimal location do not dominate the budget.

There are other considerations than cost, however. Remote locations are not only expensive, but can make it hard to service the instrument. Antarctic telescopes can only be accessed for maintenance for about three months of the year; space telescopes are, with the exception of HST, impossible to repair, service, or upgrade. A less suitable site that is within easy reach of laboratories and academic infrastructure may generate more and better science than a better, but harder to reach, site. The selection of an appropriate operating model for an instrument goes hand-in-hand with site selection, inside the overall logistics envelope for the scientific mission proposed.

As an exercise, let us consider the cost of building something like the ATCA in space or on the Moon. ATCA is an array of six 22 m antennas with a total collecting area of 2280 m^2. Presently, we are limited by rocket payload size to telescopes 15 m in diameter. This would imply that in order to get the same collecting area as the ATCA, we would need to launch at least 13 such telescopes into space. That would require several coordinated—and successful—rocket launches. It would be even more expensive to land those telescopes on the Moon, but less expensive, and easier, to position a satellite-telescope in Earth's orbit.[60] Of course, the precise costs are difficult to estimate, but certainly it would take a massive effort to establish such an instrument in space or on the Moon. Finally, one shouldn't forget that a lunar

---

[60] Less velocity change after the initial launch is required in this case.

telescope would have another type of challenge—how to communicate with Earth, given that we've isolated it from radio signals! We would need a satellite in lunar orbit to transmit information back to Earth. Space-based radio telescopes, on the other hand, while making communication with Earth much easier wouldn't be totally shielded from Earth's Radio-Frequency Interference (RFI; HDE 226868 2018). A "middleman" satellite in lunar orbit is no triviality. It adds a significant expense, and the increased potential for catastrophic failure. Modern telescopes run almost automatically, controlled by pre-prepared observing computer codes (with a human operator on-site to take manual control if needed). The more remote the instrument from Earth, the more difficult it is for mistakes to be corrected and damage to be repaired.

All these next-generation giant telescopes are very expensive. In today's (2020) money JWST costs around $10 US billion versus ~$2 billion that is the estimated costs (in 2020) of erecting a Giant Magellan Telescope (GMT), Thirty Meter Telescope (TMT), or Extremely Large Telescope (ELT). For the SKA and Advanced Telescope for High Energy Astrophysics (ATHENA) projected costs are in vicinity of $1 billion, while CTA (~$0.5 billion), and KM3NeT (~$0.5 billion) are somewhat "cheaper." The above mentioned eLISA could hit at least a $5 billion dollar price tag. Given that the International Astronomical Union (IAU) lists just under 15,000 member-astronomers world-wide, our community has a responsibility regarding the proper use of scientific resources. And finally, with the suggested launch date of 2035, Origins Space Telescope (OST) is a concept study for a far-infrared surveyor space telescope mission (for more details see https://asd.gsfc.nasa.gov/firs/).

References

Cervantes-Cota, J., Galindo-Uribarri, S., & Smoot, G. 2016, Univ, 2, 22

Cowen, D. F. 2020, JPCS, 1342, 012001

De Sabbata, V., & Weber, J. 1977, Topics in Theoretical and Experimental Gravitation Physics, Vol. 27 (Dordrecht: Reidel)

DeWitt, C. M., & Rickles, D. 2011, The Role of Gravitation in Physics: Report from the 1957 Chapel Hill Conference (Berlin: Max Planck Research Library for the History and Development of Knowledge)

Egler, R. 2006, JRASC, 100, 37

Einstein, A. 1905, AnP, 17, 891

Fegan, D. 2019, Cherenkov Reflections: Gamma-ray Imaging And The Evolution Of TeV Astronomy (Singapore: World Scientific)

Forward, R. L. 1978, PhRvD, 17, 379

Heaviside, O. 1893, Electromagnetic Theory, Vol. 1 (London: "The Electrician" Printing and Publishing)

HDE 226868, 2018, https://physics.stackexchange.com/users/56299/hde 226868, Why put a radio telescope in space instead on the Moon? Physics Stack Exchange, https://physics.stack-exchange.com/q/419938 (version: 2018-07-29)

Hulse, R. A., & Taylor, J. H. 1975, ApJL, 195, L51

Kafka, P. 1972, Are Weber's Pulses Illegal? Technical Report (Babson Park, MA: Gravity Research Foundation)

Kramer, J. 2012, A Short History of the Telescope, https://www.lcas-astronomy.org/articles/display.php?filename=a_short_history_of_the_telescope&category=beginning_astronomy

Leahy, D. A., Bianchi, L., & Postma, J. E. 2018, AJ, 156, 269

Leverenz, H., & Filipović, M. D. 2021, SerAJ, in press (doi: 10.2298/SAJ210518001L)

LIGO Scientific Collaboration, 2021, Detections https://ligo.org/detections.php

Maxwell, J. C. 1865, RSPT, 155, 459

McFadden, C. 2018, A Brief History of The Telescope: From 1608 to Gamma-Rays, https://interestingengineering.com/a-brief-history-of-the-telescope-from-1608-to-gamma-rays

Mirzoyan, R. 2014, APh, 53, 91

Newton, I. 1758, Original Letter from Isaac Newton to Richard Bentley (Normalized)

Osterbrock, D. E. 2002, James E. Keeler: Pioneer American Astrophysicist: And the Early Development of American Astrophysics (Cambridge: Cambridge Univ. Press)

Poincaré, H. 1905, CRAS, 140, 1504

Schauer, A. T. P., Drory, N., & Bromm, V. 2020, ApJ, 904, 145

Schutz, B. F. 1984, AmJPh, 52, 412

Thorne, K., & Weiss, R. 2016, A Brief History of LIGO, https://www.caltech.edu/about/news/brief-history-ligo

Völk, H. J., & Bernlöhr, K. 2009, ExA, 25, 173

Watson, F. 2004, Stargazer: The Life and Times of the Telescope (Cambridge: Da Capo)

Weber, J. 1960, PhRv, 117, 306

Weber, J. 1969, PhRvL, 22, 1320

Chapter 2

Electromagnetic Radiation

Of the messengers that bring information from the universe, *electromagnetic radiation* or *photons* dominate our understanding. A large portion of the work of multimessenger astronomy is cross-referencing the information obtained from other messengers with our basic knowledge of the universe, which is based on our observation of electromagnetic radiation. Electromagnetic radiation itself covers a vast range of phenomena, the most obvious of which is visible light. In this chapter, we provide a brief introduction to the physics of electromagnetic radiation, and define terminology to be used throughout this book.

2.1 Understanding Light—A Brief History

The drive to understand light itself was often intertwined with the drive to understand vision and the human eye in the development of science. Later in history, an interest in optical instrumentation—microscopes and telescopes—drove further research. Finally, the theory of electromagnetism provided a deep understanding of the nature of light. Modern instrumentation development combines an understanding of electromagnetic waves with research into the quantum nature of light ("quantum optics").

2.1.1 Optics—Explaining Light

The majority view in ancient Greece and the Hellenistic world (originated by Empedocles[1]) was that the eye emitted light, so that vision was akin to radar. A minority view, due to the atomists (such as Lucretius[2]) suggested that light consisted of particles emitted by the Sun. Euclid[3] and Ptolemy[4] both studied light

---

[1] Empedocles (495–435 BCE) was a Greek pre-Socratic philosopher.

[2] Titus Lucretius Carus (99–55 BCE) was a Roman poet and philosopher.

[3] Euclid of Alexandria (around mid-IV century BCE), Greek mathematician.

[4] Claudius Ptolemy (AD 100–170) was a mathematician, astronomer, geographer, and astrologer.

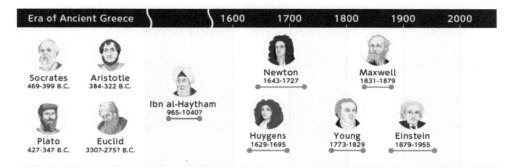

Figure 2.1. History of research on light. Image credit: Photon terrace by Hamamatsu Photonics K. K. https://photonterrace.net/en/photon/history/.

mathematically, particularly its propagation in straight lines, reflection and refraction (Figure 2.1).

The Islamic scientist Hasan ibn al-Haytham (c. 965–1040; Latinized to "Alhazen") developed a theory using geometry and anatomy, where light was emitted by all objects in all directions, and is observed when a ray of light enters the eye. He described light as a stream of particles traveling at a finite speed, explained refraction, and studied the optics of mirrors and aberrations. In Europe, Robert Grosseteste[5] and Roger Bacon[6] experimented with light and lenses. Witelo's (1230–1314) work on perception, perspective and refraction was included with al-Haytham's work in the standard 16th century optical textbook, *Opticae Thesaurus*, which constituted the state of the art prior to the Early Modern European upheavals in scientific knowledge.

Over the 17th century, two theories of light contended with each other: a wave theory and a particle theory. They also had new observations to explain: diffraction was observed by James Gregory (1638–1675)[7] and Francesco Grimaldi (1613–1663).[8] René Descartes (1596–1650) described light as a disturbance within the substance of which the entire universe was composed—the *plenum*. In Descartes' theory, light traveled faster in denser media (probably a conclusion reached by comparison with sound waves) which explains refraction. Robert Hooke (1635–1703) and Christian Huygens (1629–1695) developed Descartes' idea into wave theories of light. Huygens' theory in particular explained all the observed behavior of light—reflection, refraction, and diffraction. Because waves travel through something, Huygens proposed that light was emitted by objects, in all directions, as waves in a medium he called the *luminiferous aether*.[9]

---

[5] Robert Grosseteste (1175–1253), was a statesman, medieval philosopher, theologian, scientist, and Bishop of Lincoln.
[6] Roger Bacon OFM (1219/20–1292), was a medieval philosopher and Franciscan friar.
[7] Inventor of the Gregorian reflecting telescope.
[8] Who named the phenomenon of diffraction ("breaking up"), and instigated the practice of naming lunar craters after astronomers and physicists.
[9] In 1887, the Michelson–Morley experiment searched in vain for this aether.

Pierre Gassendi (1592–1655) published a particle theory of light, which Isaac Newton preferred over Descartes' theory. Newton published his theory of light in his book *Opticks* (1704). His theory describes light as colored particles or *corpuscles*, which could be mixed to produce colors other than those of the rainbow.[10] Mixing all colors of particles creates white light, as was demonstrated by his famous prism experiment. The theory explains the straight-line path of light, creating shadows, whereas waves do not travel purely in straight lines, bending around obstructions. In order to explain diffraction, Newton had to invoke some parts of wave theory, and he suggested that refraction was caused by an increase in the speed of light as it enters a denser medium due to the action of gravity.

Throughout the 18th century, Newton's corpuscular theory was generally accepted, although Euler argued for the wave theory. It was not until the 19th century that the dispute was settled. In 1801, Thomas Young (1773–1829)[11] published his experimental results that demonstrated the interference of light. This was clear experimental evidence that light behaved like waves, and Young also proposed that color can be explained as different wavelengths as well as the three-color explanation of human color vision.

Augustin-Jean Fresnel (1788–1827) seems not to have known of the previous wave theories, creating and presenting his own wave theory of light, which was experimentally verified by the observation of a bright spot at the center of the circular shadow cast by a disk. Further mathematical refinements to the wave theory were added by Poisson, Arago, Fraunhofer, and Airy, while new experimental work by Fizeau measured the speed of light. Léon Foucault (1819–1868)[12] was the first to measure the speed of light with enough accuracy to show that light traveled more slowly in dense media (in 1850). The wave theory of light was now unassailable, and Newton's corpuscular theory was finally abandoned.

2.1.2 Electromagnetic Radiation—The Nature of Light

The Unification of Electricity and Magnetism

At the start of the 19th century, electricity and magnetism were considered to be two separate phenomena. In 1820, Hans Christian Ørsted (1777–1851) discovered the link between electric current and magnetism, noticing the deflection of a compass needle as the current through a nearby wire was switched on or off. He determined that the magnetic field surrounding the current carrying wire was circular and perpendicular to the path of the wire. Michael Faraday (1791–1867) then discovered the inverse process, electromagnetic induction, in which an electric current is produced by a magnetic field.

In his *Treatise on Electricity and Magnetism* (1865), James Clerk Maxwell combined several known laws of electricity and magnetism: Faraday's law of

---

[10] Newton was also responsible for the establishment of the seven colors of the rainbow, by analogy with the seven notes of a musical scale; the colors indigo and orange corresponded to the semitones of the scale.

[11] The polymath who also decoded much of the Rosetta Stone, in addition to his work in medicine, linguistics, and what is now condensed matter physics.

[12] Foucault also demonstrated the rotational motion of the Earth (Foucault's pendulum) and is credited with inventing the gyroscope.

induction, which gives the electric field produced by a changing magnetic field; Gauss' law, which relates the electric field to the electric charge nearby; and the magnetic parallel to Gauss' law, which states that there are no "magnetic charges," or magnetic monopoles. He added a new law based on Ampère's law, itself an elaboration of Ørsted's work, which gives the magnetic field produced by electric currents. To the "free currents" considered by Ampère, he added a *displacement current* caused by the movement of charges in a rapidly-varying electric field. This displacement current was a theoretical insight, not detected in the laboratory work of the time. Its inclusion led to *Maxwell's equations*, a set of four equations that must be satisfied everywhere by electric and magnetic fields. Together with their supporting equations and definitions, Maxwell's equations govern electromagnetism, and were given their modern form by Oliver Heaviside. They define the relationship of the electric and magnetic fields to electrical charges and currents, and show that each field is able to generate the other.

Electromagnetic Waves

Maxwell's equations show that electric and magnetic fields can each generate the other without requiring the presence of free charges or currents. This implies that fluctuating electric and magnetic fields can propagate one another in free space. Further analysis of the equations shows that a *transverse wave* solution satisfies Maxwell's equations, and thus that an electromagnetic disturbance can create electromagnetic waves that propagate through free space.

Maxwell himself found that the known experimental data of the time showed that the speed of an electromagnetic wave was similar to the speed of light. Heinrich Hertz (1857–1894) demonstrated the production of electromagnetic waves by sparks, and showed not only that they had the same speed as light, but also could be reflected, refracted, and polarized.

In the wave solutions to Maxwell's equations, the oscillating electric and magnetic fields are in phase, perpendicular to each other, and perpendicular to the direction of wave travel. The direction of oscillation of the electric fields defines the plane of polarization of the wave. Both fields oscillate with the same frequency, which is the frequency ν (measured in hertz; Hz) of the electromagnetic wave. Because the frequency (ν) can take any value, electromagnetic radiation can have any frequency. The domain of all possible values of ν is called the *electromagnetic spectrum* (see Figure 1.1).

2.1.3 Photons—Quantized Light

At the end of the 19th century, classical physics was thought to be a complete description of the physical universe, but two experimental phenomena were proving hard to explain and both were related to light—blackbody radiation and the photoelectric effect. Blackbody radiation (Section 2.2.6) refers to the emission of electromagnetic radiation—across the entire spectrum—by any body with a non-zero temperature. The photoelectric effect allows electric current to flow through a vacuum when light falls on a metal. Neither effect could be satisfactorily explained by the known characteristics of light as described by electromagnetic waves.

The *blackbody* referred to is a body of matter that does not reflect light at all—it only absorbs light, which may then be re-emitted. Hence, the body is black. The particles that make up the body oscillate to produce electromagnetic radiation, and this is the only way that they can lose energy. So a large amount of a very tenuous gas, in which the particles very seldom collide, but the gas is still opaque, is a good example, as is the inside of a furnace. The classical arguments that lead to the formula for blackbody radiation rely on this concept of a large number of oscillators, treated as a thermodynamical system.

The Stefan–Boltzmann law, empirically derived by Josef Stefan (1835–1893) and theoretically by Ludwig Boltzmann (1844–1906), preceded Planck's blackbody law by about 20 years, stating that the total power emitted by the blackbody per unit surface area is a function only of temperature (specifically, T^4). This law, however, carries no information about the distribution of the blackbody radiation over the electromagnetic spectrum. Several attempts were made to derive a law that would give the spectral distribution of blackbody radiation. According to the Rayleigh–Jeans[13] law, based on the thermodynamic analysis of the oscillators in the blackbody, the brightness of the blackbody at a specific frequency was proportional to $\nu^2 T$. However, since the frequency can increase without limit, the power emitted from the blackbody is unlimited—this was the "ultraviolet catastrophe"—and the Rayleigh–Jeans law, though it predicted the low-frequency emission accurately, completely failed to predict the high-frequency spectrum.

The ultraviolet catastrophe was solved by Wien,[14] who used the exponential term from the Maxwell–Boltzmann distribution to reduce the high-frequency brightness, so that it would become zero in the high-frequency limit, and the total brightness would remain finite. When compared to experimental data, the Wien law reproduced the high-frequency spectral shape well, but failed to reproduce low-frequency data.

Max Planck[15] modified Wien's law to give the blackbody spectrum its final shape, reproducing experimental data accurately over the whole frequency range. Planck derived his equation to reproduce the data, and then created a rationale for it. He postulated that an oscillator with frequency ν, instead of having energy $\sim kT$ (where k is the Boltzmann constant) according to classical equipartition, could only take on energies of $nh\nu$, for integer n and Planck's constant h. As the frequency increases, these energies are more and more widely spaced, fewer and fewer of them are occupied, and the *average energy* of the oscillators at high frequency drops fast enough to prevent the ultraviolet catastrophe.

---

[13] John Strutt (1842–1919), 3rd Baron of Rayleigh (Lord Rayleigh), English physicist an Nobel prize winner in 1904. Sir James Hopwood Jeans (1877–1946) was an English physicist, astronomer, and mathematician.
[14] Wilhelm Carl Werner Otto Fritz Franz Wien (1864–1928).
[15] Max Karl Ernst Ludwig Planck (1858–1947), 1918 Nobel Laureate in physics and namesake of Germany's prestigious scientific research institutes.

The Photoelectric Effect

The photoelectric effect occurs when light falls onto a metal surface in a vacuum (Figure 2.2). If the metal surface is part of an electric circuit, connected to a negative voltage, and the positive voltage is connected to another electrode in the same vacuum vessel as the metal surface, no current will flow because the vacuum interrupts the circuit. But if light shines on the metal surface a current can flow. This can easily be explained as the energy of the light freeing electrons from the metal surface, and the electrons traveling across the vacuum to deliver a current. However, the details of the photoelectric effect were harder to explain. An electric current would only flow once the frequency of the light exceeded a certain value for that particular metal, and the kinetic energy of the freed electrons also depended on the frequency of the light, and not on its intensity. The intensity of the light only affects the magnitude of the current. These details cannot be explained by the 19th century theory of light, but could only be explained by Einstein's hypothesis that light comes in packets, with specific energy and frequency.

The Modern Understanding of Light

Einstein's insight (part of his 1905 *annus mirabilis*; miracle year) provided a physical theory to underpin Planck's blackbody radiation law. Light can be described as consisting of a stream of particles—*photons*—with discrete energies and frequencies, related by:

$$E = h\nu = \frac{hc}{\lambda} \qquad (2.1)$$

where Planck's constant, $h = 6.63 \times 10^{-34}$ J s. Photons have no mass, but do carry momentum:

$$p = \frac{h}{\lambda} = \frac{h\nu}{c}. \qquad (2.2)$$

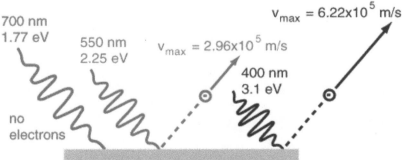

Figure 2.2. The photoelectric effect.

For a given situation, it will usually be more useful to treat electromagnetic radiation as a collection of waves or as a stream of photons. For example, when dealing with optical experiments using lenses, mirrors, filters, polarizers etc, the wave properties of light are usually used. Detection of light, on the other hand, often uses the photon concept, and our statistical treatments often assume a stream of photon events (yielding Poisson statistics). However, even in the simplest optical systems, it is common to manipulate the light by using its wave properties, and then detect it by the effect of its photons on a material. The two sets of properties cannot be separated.

2.2 The Physics of Electromagnetic Radiation

2.2.1 Wave Terminology

Frequency (usual symbol ν, but sometimes f) is a count of the number of cycles passing a point (or passing through a plane) in one second. Until 1960 this quantity was measured in units of "cycles per second" but the units were then renamed "hertz" (Hz).[16]

Wavelength (λ) is the measure, in meters, of the distance between two similar points on the wave. Radio wavelengths are in meters (m) or centimeters (cm). Microwave wavelengths are in millimeters (mm). Infrared is measured in micrometers (μm; also known as "microns"). Optical light wavelengths are on the order of a few hundred nanometers (nm) or in the non-SI units commonly used in optical astronomy, a few thousand Ångströms (Å).[17] Above the ultraviolet range, wavelength is not usually used. Then we talk about the energy of the photons (Section 2.1.3), usually in electron volts (eV). At times, λ is used as an abbreviation for the word "wavelength" and $\lambda\lambda$ for "wavelengths."

Amplitude (A) is related to the intensity or energy flux of the radiation and is measured in various ways. For use in equations such as (2.15), we define it here as the peak intensity of (the varying component of) the electric field:

$$E_0 = [\mathbf{E} \cdot \mathbf{E}]^{1/2}. \tag{2.3}$$

Or the magnetic field:

$$B_0 = [\mathbf{B} \cdot \mathbf{B}]^{1/2}. \tag{2.4}$$

The *wave velocity* (v) is measured in meters per second and is related to the frequency and wavelength by the wave formula:

$$v = \lambda\nu. \tag{2.5}$$

---

[16] Heinrich Hertz (1857–1894), German physicist who in 1888, demonstrated the existence of radio waves, as predicted in 1865 by James Clerk Maxwell, and showed that radio waves displayed reflection, refraction, interference, and polarization.

[17] The Ångström is named for Swedish physicist Anders Jonas Ångström (1814–1874), one of the founders of spectroscopy. It is equal to 10^{-10} meters. While not an SI unit, it is still used extensively in optical astronomy and in astrophysics.

This is also known as the *phase velocity* of the wave, $v = \omega/k$. The phase velocity is the speed at which any one component (or phase) of the wave (e.g., the crest of the wave) travels and it is frequency dependent.

The *group velocity* is the travel speed of any variations in the amplitude of the wave, $v_g = \partial\omega/\partial k$. The phase velocity represents the actual speed of the wave through space while the group velocity describes the speed at which energy and information carried by the wave (as variations in amplitude) travel through space. It is the group velocity that is of true importance.

2.2.2 Maxwell's equations

The Maxwell–Heaviside equations are:

$$\nabla \cdot \mathbf{D} = \rho \tag{2.6}$$

$$\nabla \cdot \mathbf{B} = 0 \tag{2.7}$$

$$\nabla \times \mathbf{E} = -\frac{\partial \mathbf{B}}{\partial t} \tag{2.8}$$

$$\nabla \times \mathbf{H} = \mathbf{j}_{\text{free}} + \frac{\partial \mathbf{D}}{\partial t}. \tag{2.9}$$

In this formulation of the equations, $\mathbf{D}$ and $\mathbf{E}$ are closely-related vector electric fields, while $\mathbf{B}$ and $\mathbf{H}$ are similarly-related magnetic fields:

$$\mathbf{D} = \varepsilon\varepsilon_0\mathbf{E} \tag{2.10}$$

$$\mathbf{H} = \frac{1}{\mu\mu_0}\mathbf{B}. \tag{2.11}$$

$\mathbf{E}$ and $\mathbf{B}$ are electric and magnetic fields in free space, while $\mathbf{D}$ (displacement field) and $\mathbf{H}$ (magnetizing field) are *derived* fields in a medium that has its own electrical and magnetic properties—the *relative permittivity* ε and *relative permeability* μ. These are multipliers of the *permittivity of free space* ε_0 and *permeability of free space* μ_0. Since the permittivities and permeabilities are scalars,[18] the vector fields $\mathbf{D}$ (the *displacement* field) and $\mathbf{H}$ (the *magnetic induction*) are parallel to their underlying vector fields $\mathbf{E}$ and $\mathbf{B}$.

$\mathbf{j}$ is the electric current density, a vector quantity, that is composed of a free current density and a magnetization current density, $\mathbf{j} = \mathbf{j}_{\text{free}} + \mathbf{j}_M$, and obeys the *continuity equation*:

$$\frac{\partial\rho}{\partial t} + \nabla \cdot \mathbf{j} = 0 \tag{2.12}$$

---

[18] In general, permittivity and permeability are tensors, but in an isotropic medium, they behave as scalars.

which can itself be derived from Maxwell's equations. The final quantity in these equations is the electric charge density ρ.

The first two equations, Equations (2.6) and (2.7), show that the electric field is generated by charges, but there is no such thing as a magnetic charge (or *magnetic monopole*). Equations (2.8) and (2.9) show that changing electric and magnetic fields produce vector magnetic and electric fields, respectively, which allows an electromagnetic disturbance to propagate in free space.

Quantities and Units The theory of electromagnetic radiation lies at the heart of our system of physical units. Because c, the speed of light in a vacuum, is a constant of nature, we use it, with the defined unit of time (second) to define the unit of distance (meter). μ_0 is also exactly defined, so ε_0 is implicitly defined.

Maxwell's equations can be written in a few different ways. They can be written purely in terms of **E** and **B** (e.g., treatment in Cottingham & Greenwood 1991), a formulation which is particularly useful if the waves are only considered in free space. They can also be written in Gaussian (cgs) units (Appendix C). In the Gaussian system, ε_0 and μ_0 are both set to 1 (e.g., see the presentation of Maxwell's equations in Rohlfs & Wilson 2004).

2.2.3 The Propagation of Electromagnetic Waves

Maxwell's equations can be used to show what happens when there are varying electric and magnetic fields due to oscillating electric charges and currents nearby. In free space (containing no charges and no currents), the **E** and **B** fields follow the equations:

$$\nabla^2 \mathbf{E} = \varepsilon_0 \mu_0 \frac{\partial^2 \mathbf{E}}{\partial t^2} \tag{2.13}$$

$$\nabla^2 \mathbf{B} = \varepsilon_0 \mu_0 \frac{\partial^2 \mathbf{B}}{\partial t^2}. \tag{2.14}$$

These are wave equations, which are satisfied by wave solutions in **E** and **B**. In the wave solutions, **E** and **B** fields are functions of time t and position **r** and take the form:

$$\mathbf{F}(\mathbf{r}, t) = F_0 \sin(\omega t - \mathbf{k} \cdot \mathbf{r}) \tag{2.15}$$

where **F** can be either **E** or **B**, F_0 is the amplitude (i.e., maximum magnitude) of the field oscillation, $\omega = 2\pi\nu$ is the *angular frequency* of the wave (and ν is the *frequency*). **k** is the *wave vector*, a vector quantity whose magnitude, k is the *angular wave number*, which is related to the number of waves per unit length ($k = 2\pi/\lambda$).[19] The scalar product $\mathbf{k} \cdot \mathbf{r}$ shows how the wave solution varies as the vector position **r** aligns with the wave vector **k** that points in the direction of travel of the wave.

---

[19] You might also meet the *wave number*, which is the number of oscillations per unit length, usually in units of cm^{-1}. It's traditionally used in the infrared to microwave spectral range.

By comparison with the general form of a wave equation, these wave solutions travel at speed c:

$$c = \sqrt{\frac{1}{\varepsilon_0 \mu_0}} \qquad (2.16)$$

in free space.

Electromagnetic Waves in Isotropic Media
The *electric displacement* vector field $\mathbf{D}$ used in Maxwell's equations comprises the electric field $\mathbf{E}$ and the *polarization field*:

$$\mathbf{D} = \varepsilon_0 \mathbf{E} + \mathbf{P}. \qquad (2.17)$$

$\mathbf{P}$ arises from the effect of the electric field on the bound electric charges (i.e., atoms) that fill the space of a dielectric material (crudely, an insulator). Electrons are shifted to one side of their nuclei by the electric field, but remain bound, and this contributes more components to the electric field, current, and charge.

In the case of electromagnetic waves traveling through a medium (i.e., not through free space), the oscillating electric field will produce a polarization field. If the medium is *isotropic*,[20] the polarization field will be parallel to the electric field (see Equation (2.10)) and so:

$$\mathbf{D} = \varepsilon \varepsilon_0 \mathbf{E} \qquad (2.18)$$

where ε is the *relative permittivity* or *dielectric constant* of the medium.[21]

In the general (i.e., non-isotropic) case, the tensor $\underline{\varepsilon}$ maps $\mathbf{E}$ to $\mathbf{D}$, but for isotropic media, the tensor becomes a frequency-dependent scalar $\varepsilon(\nu)$. The magnetic field $\mathbf{H}$ is related to $\mathbf{B}$ in the same way:

$$\mathbf{B} = \mu_0 \mu \mathbf{H} \qquad (2.19)$$

where μ and μ_0 are the *relative permeability* of the material and the *permeability of free space*.

In the case of an electromagnetic wave traveling through an isotropic medium, the wave speed is given by:

$$u = \sqrt{\frac{1}{\varepsilon \varepsilon_0 \mu \mu_0}} = c \sqrt{\frac{1}{\varepsilon \mu}}. \qquad (2.20)$$

The *refractive index* $n = c/u$ is therefore:

$$n = \sqrt{\frac{1}{\varepsilon \mu}} \qquad (2.21)$$

---

[20] The most common case of a non-isotropic medium is the phenomenon of *birefringence*.
[21] ε_0 is the *permittivity of free space*.

and in most cases, the medium is non-magnetic, $\mu \approx 1$, $n \approx \sqrt{\varepsilon}$. The refractive index is the factor by which the medium slows down electromagnetic waves.

Electromagnetic Waves in Plasma

The macroscopic polarization field (Equation (2.17)) is the total electric dipole moment per unit volume of dielectric material. It can also be treated as the sum of many microscopic dipoles throughout the volume. If there are N dipoles per unit volume, each with charge q and separation $\mathbf{d}$:

$$\mathbf{P} = Nq\mathbf{d}. \tag{2.22}$$

In an astrophysical plasma, electrons and ions coexist, but rarely collide due to the low density of the medium. If an electromagnetic wave passes through the plasma, the oscillating electric field will cause electrons and ions to move in opposite directions, producing microscopic electric dipoles. As with the dielectric case above, these dipoles generate a macroscopic polarization field. The separation of the dipoles is:

$$\mathbf{d} = -\frac{e}{m_e\omega^2}\mathbf{E}, \tag{2.23}$$

where e is the charge of the electron and m_e is its mass. So:

$$\mathbf{P} = -\frac{Ne^2}{m_e\omega^2}\mathbf{E}. \tag{2.24}$$

From Equations (2.17) and (2.18):

$$\mathbf{P} = \varepsilon_0\mathbf{E}(\varepsilon - 1) \tag{2.25}$$

and so:

$$\varepsilon = 1 - \frac{Ne^2}{\varepsilon_0 m_e\omega^2}. \tag{2.26}$$

The *plasma frequency*, ω_p, is defined as:

$$\omega_p^2 = \frac{Ne^2}{\varepsilon_0 m_e} \tag{2.27}$$

and so:

$$\varepsilon = 1 - \frac{\omega_p^2}{\omega^2} \tag{2.28}$$

and the refractive index:

$$n = \sqrt{1 - \frac{\omega_p^2}{\omega^2}}. \tag{2.29}$$

At low frequencies $\omega < \omega_p$, n becomes imaginary, and the plasma reflects electromagnetic radiation rather than propagating it. At high frequency, the plasma propagates electromagnetic radiation, but with steadily increasing refractive index.[22]

The most obvious example of this behavior is the ionosphere of the Earth, which blocks extraterrestrial radio waves at low frequencies and bounces terrestrial radio transmissions over the horizon, as used in shortwave radio. The plasma frequency of the ionosphere depends on its electron density, which in turn depends on the Sun. Converting from angular frequency ω to the more common frequency ν, the plasma frequency is:

$$\nu_p = 8.97\sqrt{N} \text{ kHz} \qquad (2.30)$$

where N is in units of particles per cubic centimeter (cm^{-3}).

For $\omega > \omega_p$, the wave propagates through the plasma with group velocity:

$$v_g = c\left(1 + \frac{\omega_p^2}{\omega^2 - \omega_p^2}\right)^{-1/2} \qquad (2.31)$$

so the speed of propagation is frequency dependent. The travel time of the radiation is D/v_g, where D is the distance between source and receiver. The frequency dependence of the travel time is:

$$\frac{d(D/v_g)}{d\omega} \approx -\frac{D}{c}\frac{\omega_p^2}{\omega^3} \qquad \text{with} \qquad \omega \gg \omega_p. \qquad (2.32)$$

In the case of constant radiation, this will not be obvious, but if there is temporal structure in the radiation, such as the pulse from a pulsar, or a fast radio burst, there will be a frequency dependence in the arrival times of the pulse or burst. This is termed *frequency dispersion*.

Frequency dispersion increases with the distance traveled by the light through a plasma and (from Equations (2.32) and (2.27)) with the electron number density n.

Because we can only see the cumulative effect of all the plasmas encountered along the line of sight, we can only estimate the electron density integrated along the line of sight to the source, or *dispersion measure*:

$$\int_0^D n(s)ds = D\langle n \rangle \qquad (2.33)$$

where $\langle n \rangle$ is the average electron density.

If most of the electrons are found around the object rather than in the interstellar medium, these equations yield an estimate of the local electron density at the source. Conversely, multiple dispersion measurements can be used to build up a picture of the electron density distribution of the interstellar medium.

---

[22] $n < 1$ implies that the *phase velocity* of the electromagnetic wave is greater than c, but the *group velocity* is less than c.

Energy, Momentum, and Pressure

Energy Density, *U*: The energy density (i.e., energy per unit volume) of electro-magnetic fields is given by

$$U = \frac{1}{2}(\mathbf{E} \cdot \mathbf{D} + \mathbf{B} \cdot \mathbf{H}) = \frac{\varepsilon_0}{2}E^2 + \frac{\mu_0}{2}H^2 \qquad (2.34)$$

where *E* and *H* are the magnitudes of the electric and magnetic fields, **E** and **H**, giving the scalar quantity *U*.

Although the energy density of the fields associated with the electromagnetic radiation is a scalar quantity, electromagnetic radiation is able to transport energy, so it is possible to define a vector quantity. This quantity, the energy flowing across a unit area per unit time, is given by the *Poynting vector*[23] **N**:

$$\mathbf{N} = \mathbf{E} \times \mathbf{H} \qquad (2.35)$$

with the vector pointing in the direction of wave propagation.

Momentum and Radiation Pressure: Electromagnetic radiation also carries momentum in its direction of propagation. From the quantum theory of light, a photon has energy $E = h\nu$ and from the theory of special relativity, it has momentum $p = E/c = h\nu/c$. The momentum crossing a surface per unit area per unit time can therefore be derived from the Poynting vector, $\mathbf{N}/c$.

Momentum transfer per unit area per unit time is equivalent to force per unit area, or pressure. Thus, electromagnetic radiation exerts a pressure on a surface—as photons are absorbed, their momentum is given up to the surface, and a pressure of $\mathbf{N}/c$ is applied. If the photons are reflected, they lose their momentum, and gain an equal and opposite momentum, so the momentum transfer is twice as high, and the radiation pressure is $2\mathbf{N}/c$.

2.2.4 Polarization

Electromagnetic waves have spatial structure—the electric and magnetic fields, while varying in strength, lie in specific directions, perpendicular to one another and to the direction of energy flow. The direction of the fields can change over space and time. *Polarization* is the way we describe this spatio-temporal structure.

The behavior of the electric field is used to define polarization. Imagine a plane in space, with an electromagnetic wave traveling perpendicular to it. The electric field will be a vector lying within that plane, whose magnitude is the field strength, and the orientation of the electric field vector with time defines the polarization state of the light.

Strictly speaking, all light is polarized. Under normal conditions, light is randomly polarized—when you measure the electric field, its direction will vary randomly—but we usually call this *unpolarized* light. Polarization generally refers to light with easily-defined spatio-temporal structure of the electric field. The general case for polarization is *elliptical polarization*. In the plane intersecting the electromagnetic wave imagined above, the electric field vector of elliptically polarized light changes magnitude and

---

[23] Named after the English physicist, John Henry Poynting (1852–1914).

direction with time, seeming to rotate around an ellipse in the plane (so that its magnitude is greater in some directions than others). Elliptical polarization includes the two limiting cases, *linear polarization* (Figure 2.3) and *circular polarization*.

In the case of linear polarization, the ellipse in the plane collapses to a straight line, and the electric field oscillates in one direction only, with no change over time. For circularly polarized light, the electric field vector rotates in a circle, so the electric field strength remains constant over time, with varying direction.

The different forms of polarization are interconvertible, being different descriptions of the same physical phenomena. Light can generally be described as a combination of two linear polarized components, with the components having electric fields perpendicular to each other. If the two components are equal in magnitude with the correct phase relationship, the light is circularly polarized. Depending on the phase relationship, this circular polarization can be "left" or "right" circularly polarized (LCP or RCP; Figure 2.4). Linear polarization components with different magnitudes produce elliptical polarization. In the same way, a superposition of LCP and RCP light

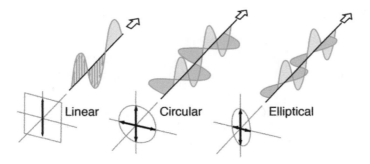

Figure 2.3. A plane polarized electromagnetic wave. The arrow shows E_o, the magnitude of the electric component of the wave, in the plane of polarization. Image credit: HyperPhysics/R. Nave.

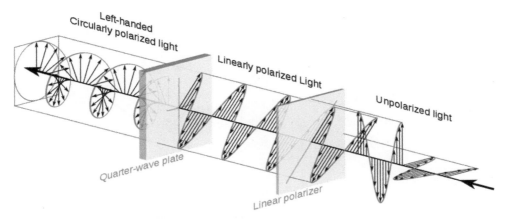

Figure 2.4. An elliptically (circular) polarized electromagnetic wave. The arrows show how **E**, the electric component of the wave, is rotating around the directional path of the wave. To improve clarity, the magnetic component, with **B** always at 90° to **E**, is not shown. Image credit: Wikimedia (Dave3457).

produces linearly-polarized light. This has particular importance in the coherent detection of electromagnetic radiation (e.g., radio or submillimeter-wave detection) in which the receiving element is generally sensitive to a specific type of polarization, such as linear polarization in a specific direction, RCP or LCP.

Faraday Rotation

When an electromagnetic wave passes through an ionized medium which is threaded by a magnetic field, the plane of polarization of that wave is twisted by *Faraday rotation* (Figure 2.5). For unpolarized light, there will be no overall effect, but if the light is polarized, that polarization will be affected. From a physical perspective, this can be seen as follows: the electric field of the electromagnetic wave causes the electrons in the plasma to oscillate; but the moving electrons experience a force due to the magnetic field threading the plasma. This force will push the electrons' plane of oscillation, twisting it.

For a more mathematical approach, we treat the linearly-polarized electromagnetic wave as the superposition of the two circular polarizations, RCP and LCP. The dielectric constant ε of a magnetized plasma is:

$$\varepsilon = 1 - \frac{4\pi m e^2}{m_e \omega (\omega \pm \omega_c)} \tag{2.36}$$

using the + for LCP and − for RCP, and where the electron *gyro-frequency* or *cyclotron frequency* is:

$$\omega_c = \frac{eB}{m_e c}. \tag{2.37}$$

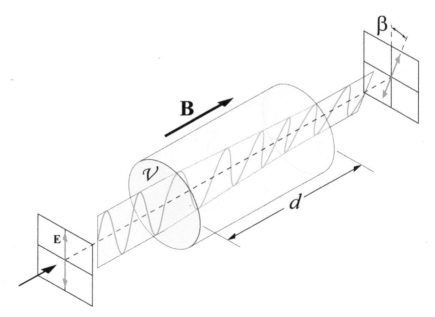

Figure 2.5. Polarization rotation due to the Faraday effect. Image credit: Wikimedia (CC BY-SA 3.0).

Differing dielectric constants for RCP and LCP waves yield different propagation speeds. Since one polarization will lead the other, the overall effect is to rotate the plane of (linear) polarization.

Faraday rotation is far more significant for a magnetic field parallel to the propagation of the wave, and so measurement of Faraday rotation effectively probes only the line-of-sight magnetic field $B_\parallel$. The change of polarization direction is:

$$\Delta\psi \propto \int_0^L B_\parallel(z)N_e(z)\mathrm{d}z \qquad (2.38)$$

where the EM wave is traveling in the z-direction along a path length L in which both magnetic field and electron density vary with z.

Faraday rotation is an invaluable probe of the line-of-sight magnetic field, but interpretation of such measurements is difficult. The original plane of polarization is unlikely to be known, and the effect depends on electron density as well as magnetic field. Pulsar radiation can overcome these drawbacks. For example, the pulse dispersion also depends on electron density, and the change in polarization angle can be tracked in frequency, since the pulse's frequency structure is well-known.

Because the waves suffering Faraday rotation must also be affected by frequency dispersion, the electron density cross-section of the ISM can also be studied by observations of the Faraday rotation of the radio pulses from pulsars. If all frequencies emitted in the pulse have the same initial polarization plane, we should see a different Faraday rotation for each frequency, which is in turn, dependent on the ISM's electron density cross-section.

2.2.5 Geometric Optics: Huygens–Fresnel and Fermat's Principles

Although the physical nature of light can be shown to be wavelike, it is very often possible to treat light as a *ray*—that is, a phenomenon that propagates in straight lines without an inherent tendency to spread out. This is the basis of *geometric optics* in which the behavior of light in an optical system is analyzed by *ray-tracing*: light travels between *optical elements* (e.g., lenses, mirrors, prisms, etc) in straight lines, and these optical elements then affect the light in specific ways. The resulting *ray diagram* shows how the light propagates through the system.

Fermat's principle is fundamental to the use of geometric optics, stating that light always travels on the shortest path (minimizing the time) between two points. This allows the behavior of light to be predicted: light will usually travel in straight lines, but the shortest-time path through a refractive optical element (e.g., a lens) will deviate from a straight line. Those who argued against this statement were troubled by the appearance of light having the intelligence to discern the shortest path between two points. This objection is overcome by the *Huygens–Fresnel principle*, which states that every point on a wavefront is the source of a new set of waves. The principle is used to explain the diffraction of light (or any wave) around corners or edges. Fermat's principle can be proven using the Huygens–Fresnel principle and the concept of interference, expressing geometric optics as a valid approximation to the results of the wave theory of light.

Geometric optics is generally most useful when the optical system is very large compared to the wavelength of light used. Thus, even highly-sophisticated modern visible-light optical telescopes and instruments are often designed using geometric optics (expressed in ray-tracing software), while radio telescopes often need to be designed around electromagnetic wave theory.

2.2.6 Blackbody Radiation

Material bodies emit electromagnetic radiation. In the case of a blackbody (see Section 2.1.3), the spectral distribution of the emitted radiation depends only on the temperature of the body, while the amount of radiation depends on temperature and surface area.

According to Planck's radiation law, the power radiated from the blackbody per unit emitting surface area into a steradian of solid angle at frequency ν, over a frequency interval $d\nu$:

$$B_{\nu}(T) = \frac{2h\nu^3}{c^2} \frac{1}{e^{h\nu/kT} - 1}. \tag{2.39}$$

The Wien approximation can be recovered from the Planck law by setting the frequency to be very high ($h\nu \gg kT$):

$$B_{\nu,\mathrm{W}}(T) = \frac{2h\nu^3}{c^2} e^{-\frac{h\nu}{kT}}. \tag{2.40}$$

Similarly, the Rayleigh–Jeans approximation can be derived by setting frequency to be low ($h\nu \ll kT$):

$$B_{\nu,\mathrm{RJ}}(T) = \frac{2\nu^2 kT}{c^2}. \tag{2.41}$$

These equations have equivalents for the power radiated per unit emitting surface area into a steradian at wavelength λ, over a wavelength interval $d\lambda$:

$$B_{\lambda}(T) = \frac{2hc^2}{\lambda^5} \frac{1}{e^{hc/\lambda kT} - 1} \tag{2.42}$$

$$B_{\lambda,\mathrm{W}}(T) = \frac{2hc^2}{\lambda^5} e^{-\frac{hc}{\lambda kT}} \tag{2.43}$$

$$B_{\lambda,\mathrm{RJ}}(T) = \frac{2ckT}{\lambda^4}. \tag{2.44}$$

These formulations of Planck's law can be integrated over either frequency or wavelength to obtain the Stefan–Boltzmann law, which gives the power radiated per unit area of the body's surface, integrated over the whole spectrum:

$$B(T) = \sigma T^4 \tag{2.45}$$

in which the *Stefan–Boltzmann constant* is:

$$\sigma = \frac{2\pi^5 k^4}{15 c^2 h^3} = 5.67 \times 10^{-8}\ \text{W m}^{-2}\ \text{K}^{-4} = 5.67 \times 10^{-5}\ \text{erg s}^{-1}\ \text{cm}^{-2}\ \text{K}^{-4}.$$

The Planck functions (Equations (2.39) and (2.42)) can be used to find the peak of the blackbody emission. The peak depends only on temperature, and is given by Wien's displacement law:

$$\nu_{\max} = (5.8789 \times 10^{10}\ \text{Hz})(T/\text{K}) \tag{2.46}$$

$$\lambda_{\max} = \frac{2.8978 \times 10^{-3}\ \text{m}}{T/\text{K}}. \tag{2.47}$$

If the peak of the spectrum is calculated in frequency ($\nu_{\max}$) and this frequency is converted to wavelength, the resulting wavelength will not be the same as the spectral peak calculated in wavelength $\lambda_{\max}$. This rather counter-intuitive result arises from the fact that the two Planck functions are density functions, and the change from unit frequency to unit wavelength shifts the peak.

The Wien displacement law explains the shift in color of a glowing object (such as a star) as its temperature rises (see Figure 2.6). A hotter object will emit most of its light at shorter wavelengths, so its overall color will move from red to orange to

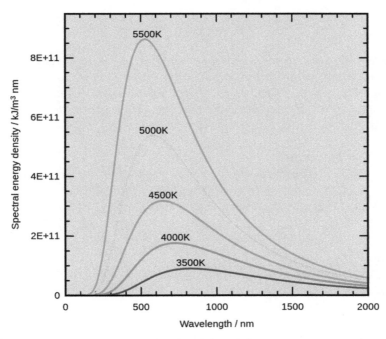

Figure 2.6. Blackbody emission curves, showing the relationship between temperature and energy emission. This illustrates Wien's displacement law, and also shows that a hotter body radiates more *at every wavelength*. The total energy emitted for a given temperature is proportional to the area under the curve for that temperature.

yellow with increasing temperature. Even hotter objects generally look white, shifting from a greenish white to a bluish white as the temperature increases. The reason that these objects generally look white is that there is still very intense emission at longer wavelengths, and so very hot objects do not look blue, even if their blackbody spectral peak is in the blue range. Objects hotter than about 10,000 K will have their blackbody peak outside the visible range of wavelengths (in the ultraviolet), while relatively cool bodies ($\leqslant$ 2000 K) will mainly emit in infrared wavelengths, or the even longer submillimeter and millimeter wavelengths.

Brightness Temperature

At any given frequency or wavelength, the brightness of the blackbody rises monotonically with temperature. This implies that, even though the emission peak of a hotter body is in a bluer part of the spectrum than that of a cooler body, the hotter body is still brighter than the cooler body *at the frequency or wavelength of peak emission of the cooler body*.

A further implication is that the temperature can be used as a proxy for brightness. Assume that the brightness of a celestial object is found by measuring the power intercepted by an instrument, integrated over a small solid angle of the sky (the "beam"), and that this whole beam terminates on a blackbody of uniform temperature. When combined with the Rayleigh–Jeans approximation, this allows the *brightness temperature T_b* to be defined:

$$T_b = \frac{\lambda^2}{2k} B_\nu. \tag{2.48}$$

If the Rayleigh–Jeans approximation is valid (i.e., at low frequency) and the beam is indeed pointed toward blackbody radiation, then T_b is equal to the thermodynamic temperature of the blackbody. If these are not true, T_b may be a completely invalid estimate of the true temperature, but, being proportional to B_ν, it can still be, and often is, used as a pure brightness measurement.

Emissivity and Absorptivity

The blackbody is a theoretical concept; real objects are not perfect black bodies, though they may approximate them. The brightness of a real object at temperature T, $I_\nu(T)$, is related to the blackbody emission at that temperature by the *emissivity, ε_ν*:

$$I_\nu = \varepsilon_\nu B_\nu(T) = \varepsilon_\nu \frac{2h\nu^3}{c^2} \frac{1}{\exp(h\nu/kT) - 1}. \tag{2.49}$$

In general, the emissivity will depend on frequency, and may depend on many other factors too. The *absorptivity κ_ν* is the fraction of incident radiation power at a given frequency that is absorbed rather than reflected. Emissivity and absorptivity are related by Kirchoff's Law:

$$\frac{\varepsilon_\nu}{\kappa_\nu} = B_\nu(T). \tag{2.50}$$

This is often phrased as "good emitters are good absorbers," but it also implies that a good absorber at a given frequency is also a good emitter *at that frequency*.

Blackbody Radiation in the Universe
Most stars deviate significantly from a blackbody spectrum, particularly cooler ones. The *effective temperature* T_{eff} is defined by comparing the power output of the star to that of a blackbody, using the Stefan–Boltzmann law. The power output of the star per unit emitting surface area over the whole spectrum:

$$\mathcal{F}_{\text{bol}} = \sigma T_{\text{eff}}^4 \qquad (2.51)$$

and if the star is taken to be spherical, the luminosity (i.e., power output) of the star:

$$L_* = 4\pi R^2 \sigma T_{\text{eff}}^4 \qquad (2.52)$$

where R is the stellar radius. The effective temperature of the Sun is around 5780 K, but the solar spectrum deviates somewhat from the spectrum of a blackbody with this temperature.

Dense gas clouds also emit thermal radiation from the dust grains embedded in the gas. The emissivity of these grains at long wavelengths varies smoothly with frequency (usually modeled as a power-law), and produces a modified blackbody spectrum.[24]

The best example of a natural blackbody radiator is the universe itself—the cosmic microwave background precisely follows the blackbody spectrum of a body with temperature 2.72548 ± 0.00057 K (Fixsen 2009).

2.3 The Electromagnetic Spectrum

While the electromagnetic spectrum is continuous, it is usually subdivided for convenience. The subdivisions correspond to the emission mechanisms that generate the radiation, the means of detecting it, or the spectral range allowed through the Earth's atmosphere (Figure 1.1; Appendix F). The different named spectral regions are defined by frequency, energy or wavelength, and can overlap each other.

Frequencies and wavelengths are generally given in SI and cgs units (Appendices B and C), i.e., Hz and m, with suitable prefixes (Appendix A). Wavelengths are sometimes given in the non-SI Ångström (1 Å = 0.1 nm = 10^{-10} m; Appendix D). Energies are often used at the highest-frequency end of the spectrum, generally given in the non-SI electron volts, eV. The electron-volt is the kinetic energy gained or lost by an electron as it moves through a potential difference of one volt. 1 eV = 1.60×10^{-19} J = 1.60×10^{-12} erg.

---

[24] This spectrum is sometimes referred to as a "gray body" in the literature, even though the frequency dependence of the emissivity arguably produces an additional "color." A gray body is also defined as one that has a constant emissivity, without frequency dependence. The term is best avoided.

The Radio to Terahertz Spectrum

For astronomical purposes, the low-frequency end of the electromagnetic spectrum is defined by the Earth's ionospheric cutoff, which varies between about 10–30 MHz, depending on ionospheric conditions. The atmosphere is transparent to radiation between this cutoff and about 26 GHz, or 1.2 cm wavelength, and this is generally considered to be the radio region of the spectrum. Radio emission largely comes from charged astrophysical plasmas (through synchrotron or bremsstrahlung radiation; Chapter 6) and from the 21 cm wavelength neutral hydrogen spin-flip transition.

Higher frequencies, about 30–300 GHz are often known as the millimeter-wave (or mm-wave) spectrum, with wavelengths of about 1–10 mm, mainly seen through atmospheric windows at around 7 mm, 3 mm and 1 mm. The submillimeter-wave (submm) spectrum (300 GHz–1 THz) is largely observed through the 850 μm, 450 μm, and 350 μm (350 GHz, 450 GHz, and 800 GHz) windows. Emission in this region of the spectrum is dominated by the thermal (modified blackbody) radiation of interstellar dust, and the rotational quantum transitions of interstellar molecules (see Appendix F).

The terahertz spectrum (from 1 THz to about 3 THz) is almost inaccessible from the ground, but observations are occasionally made from very high, dry sites. The emission arises from interstellar dust (as for the submm) and from fine-structure quantum transitions of atomic and ionized gas in the interstellar medium. The terahertz spectrum overlaps with the far-infrared region, with wavelengths of order 100 μm.

Infrared, Visible, and Ultraviolet Light

At frequencies higher than about a terahertz, wavelength is the more commonly used to define the spectrum. The infrared (IR) region stretches from the poorly-defined long-wavelength end around 1–2 THz (150–300 μm) to the red limit of human vision, around 700 nm (though some definitions consider the IR to extend all the way to $\lambda \sim 1$ mm). The near-IR, about 700 nm to 5 μm wavelength, corresponds to a series of atmospheric windows. Wavelengths longer than 5 μm are called mid-IR. The transition between mid-IR and far-IR is often taken to be 25 μm, where the atmosphere becomes fully opaque all the way to the THz and submm windows (see Appendix F).

Visible light is generally taken to cover wavelengths from about 400 nm to about 700 nm. The typical colors perceived by humans are violet (~400 nm), blue (~450 nm), cyan (~500 nm), green (~550 nm), yellow (~580 nm), orange (~600 nm), and red (~700 nm).

The ultraviolet spectrum is generally taken to start at the atmospheric ultraviolet cutoff around 320 nm and to end at 10 nm. At 91.2 nm, the *Lyman limit* lies within this range. This is the wavelength of photons, equivalent to an energy of 13.6 eV, that will ionize a hydrogen atom in its ground state. The spectral range between 91.2 nm and 10 nm is known as the extreme ultraviolet (EUV), and these photons will be absorbed by the neutral hydrogen in the interstellar medium. The EUV therefore probes only a small volume of space around us (the "Local Bubble").

The High-energy Spectrum

The high-energy electromagnetic spectrum, comprising X-rays and gamma-rays, is generally defined in energy (eV or more often keV). The short-wave end of the extreme ultraviolet, about 10 nm, is equivalent to an energy of about 120 eV. The soft X-ray region extends from 120 eV to about 10 keV, and 10 keV to 100 keV photons are hard X-rays (see Appendix F).

Photons with energies above about 100 keV are known as soft gamma-rays. The transition between soft and hard gamma-rays is unclear, but is generally placed in the MeV range. Hard gamma-rays then extend in energy all the way up to about 1 TeV, after which they are called "very high energy" (VHE) gamma-rays.

2.4 Further Reading

A historical treatment of blackbody radiation presenting the development of the idea from the classical treatment to the Planck law, is given in the Feynman Lectures on Physics (Feynman et al. 2010), Chapter 41. Marr et al. (2015) explain the law by consideration of the Bose–Einstein statistics of a photon gas, which is anachronistic but easier to understand. Wilson et al. (2018) provide a clear account of the basic results, without going into much detail on the physics.

References

Cottingham, W. N., & Greenwood, D. A. 1991, Electricity and Magnetism (Cambridge: Cambridge Univ. Press)

Feynman, R. P., Leighton, R. B., & Sands, M. 2010, The Feynman Lectures on Physics (New millennium ed.; New York, NY: Basic Books)

Fixsen, D. J. 2009, ApJ, 707, 916

Marr, J. M., Snell, R. L., & Kurtz, S. E. 2015, Fundamentals of Radio Astronomy: Observational Methods (London: Taylor & Francis)

Rohlfs, K., & Wilson, T. L. 2004, Tools of Radio Astronomy (Berlin: Springer)

Wilson, T. L., Rohlfs, K., & Hüttemeister, S. 2018, Tools of Radio Astronomy—Set: 6th Edition with Problems and Solutions Book (Berlin: Springer)

Chapter 3

The Measurement of Cosmic Messengers

Electromagnetic waves transport energy from one place to another, generally radiating out from a central source. The measurement of electromagnetic radiation therefore allows us to estimate the amount of energy radiated from that source, and hence its physical characteristics. These measurements and estimates are the foundation of modern astrophysics.

3.1 The Measurement of Electromagnetic Radiation

3.1.1 Measured Quantities for Electromagnetic Radiation

There are many different quantities associated with electromagnetic radiation that can be measured, each with their own units. The appropriate quantity to measure depends on the measurement technique and the physics of the emitting object.

Specific Intensity, I_ν
We seek to measure the amount of electromagnetic radiation coming to us from an astrophysical source by measuring the amount of energy that crosses some defined surface. This surface may be thought of in practical terms as the illuminated face of a detector such as a charge-coupled device (CCD), but it is a purely theoretical construct.

　　If we observe the same astrophysical radiation for twice as long, twice as much energy will pass through the surface; if the surface has twice the area, twice as much radiation will pass through. Radiation can come to the surface from any direction, but the surface will only measure radiation from a particular range of directions (in the case of a CCD, this would be defined by the telescope in front of it, which "looks" in a particular direction). If the surface accepts radiation from twice as large a range of directions (i.e., twice the solid angle), twice as much radiation will pass through. The amount of radiation passing through the surface will be smaller, though, if this range of directions is at an angle to the direction normal to the surface—the effective area of the surface will be reduced. Finally, only radiation over a given frequency range is measured. If this frequency range is reduced to half its previous value, half as much radiation will be measured.

So, overall, the amount of energy passing through a surface due to incoming astrophysical electromagnetic radiation is proportional to the time interval of measurement, the area of the surface, the solid angle which that area can see, and the frequency range that can be measured. It is also reduced by a geometric factor depending on the angle between the direction of the radiation that is seen and the normal to the surface. The situation is sketched in Figure 3.1: the area of the detector surface is dA, and the radiation is coming in from directions in a solid angle $d\Omega$, which is centered on the direction (θ, ϕ) in polar coordinates. θ is the angle between this direction and the normal to the surface, so the effective receiving area is $dA \cos \theta$. Radiation is measured over a frequency range $d\nu$ and over a time interval dt. Even with all these effects taken into account, some sources of radiation deliver more energy than others, and we describe this by the *specific intensity* $I_\nu(\theta, \phi)$, the flow of energy through a unit surface per unit time, per unit solid angle, per unit frequency, in a specific direction:

$$dE = I_\nu(\theta, \phi)\, dA \cos \theta\, dt\, d\Omega\, d\nu. \tag{3.1}$$

If the radiation source is spatially resolved, so that all of the rays in the solid angle $d\Omega$ come from the astrophysical source, *the specific intensity is independent of the source distance*. This result arises from the combination of area and solid angle in the definition: If a source is twice as far away, its radiation will be spread out over four times the area, so the same receiving area dA intercepts a quarter of the radiation. But the solid angle $d\Omega$ remains the same, and the rays in $d\Omega$ come from four times the area of the source, so there is four times as much radiation. The two effects

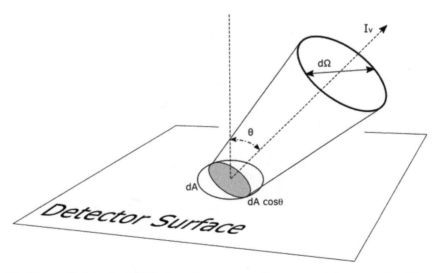

Figure 3.1. The specific intensity I_ν, is the flow of energy per unit area per unit time, per unit solid angle, per unit frequency, in a specific direction.

exactly cancel out. The specific intensity is therefore a physical property of the resolved, radiating source.[1]

Specific intensity is defined as a monochromatic quantity I_ν in units of erg cm^{-2} s^{-1} sr^{-1} Hz^{-1} (CGS)[2] or W m^{-2} sr^{-1} Hz^{-1} (SI) (Appendix B, C, and D). It can also be defined as a monochromatic quantity in wavelength rather than frequency, I_λ. I_λ has units like erg cm^{-2} s^{-1} sr^{-1} Å^{-1} or W m^{-2} sr^{-1} nm^{-1}. The quantities can be converted, e.g.:

$$I_\nu = 3.33 \times 10^{-19} \lambda^2 I_\lambda \tag{3.2}$$

where the wavelength is measured in Ångströms.

In many cases, the radiation will be measured by a broadband instrument, so that the quantity estimated will be the specific intensity integrated over a specified frequency range or the whole electromagnetic spectrum. In this case, it will have units of erg cm^{-2} s^{-1} sr^{-1} or W m^{-2} sr^{-1}.

Flux Density, F_ν

Many sources of astrophysical electromagnetic radiation are *compact*—their angular structure seen from Earth is smaller than the solid angle $d\Omega$ used to observe them. This implies that, if twice the solid angle is used to observe the source, no more radiation will be measured, because all the radiation from that source that can be intercepted by the surface is already being intercepted.

The same idea can be applied to sources that are larger than $d\Omega$—the specific intensity can be integrated over solid angle, so that all radiation from it is captured. This allows us to define the *flux density F_ν* (often denoted S_ν in radio astronomy):

$$F_\nu = \int_S I_\nu \cos\theta \, d\Omega \tag{3.3}$$

$$dE = F_\nu \, dA \, dt \, d\nu. \tag{3.4}$$

By integrating over the solid angle, the effect of θ in reducing the effective area of the detecting surface is taken into account, and the physical area may be used to convert flux density to energy. The dependence on direction (θ, ϕ) is also removed, so the flux density is a property of the astrophysical source, rather than depending on the direction in which you're looking.

Because the flux density measures the net flow of energy across the surface, in the case of completely isotropic emission (from both sides of the surface), the net energy flow, and hence flux density, will be zero. This quantity is therefore useful to measure the apparent brightness of an astrophysical source such as a star or galaxy.

The flux density can be given either over a unit frequency interval (F_ν in erg cm^{-2} s^{-1} Hz^{-1} or W m^{-2} Hz^{-2}) or a unit wavelength interval (F_λ in erg cm^{-2} s^{-1} Å^{-1} or W m^{-3}).

---

[1] However, there must be no absorption of the light between the source and observer. This is not generally the case—a more distant source is probably attenuated more than a nearby one.
[2] The use of CGS units like this is still widespread in the astronomical literature, and is even more widespread in historical papers.

It can be integrated over frequencies to give a *flux F* in erg cm^{-2} s^{-1} or W m^{-2}, either over the whole spectrum or (more commonly) over a defined band.

For most situations, the physical units used above are far too large for the tiny amounts of energy deposited by astrophysical electromagnetic radiation. A more useful unit, the jansky (Jy), has been defined to be 10^{-26} W m^{-2} Hz^{-1} (SI) or 10^{-23} erg cm^{-2} s^{-1} Hz^{-1} (CGS). This is the main flux density unit in radio astronomy, and is sometimes used in other wave bands as well[3] (Appendix B, C, and D).

Because flux density is derived from specific intensity by integrating over solid angle, the flux density is distance-dependent: a more distant source will have a lower flux density than a nearby one. The flux density of a distant source of radius R and specific intensity I_ν can be approximated by:

$$F_\nu \approx I_\nu \frac{R^2}{D^2} \tag{3.5}$$

for a small source, so that $\cos \theta \approx 1$ and $d\Omega$ is small.

Mean Intensity, J_ν

The specific intensity is a way to quantify the amount of radiation propagating through free space toward an observer from a distant object. However, sometimes we are more interested in the amount of radiation propagating within an astrophysical system—such as when studying the astrophysics of nebulae. For this purpose, the *mean intensity J_ν* is more useful. Because the radiation is coming in from all directions, we average over all directions:

$$J_\nu = \frac{1}{4\pi} \iint I_\nu(\theta, \phi) \, d\theta \, d\phi. \tag{3.6}$$

Since this is averaged over 4π steradians, the total energy entering a region per unit area per unit time is $4\pi J_\nu$, with units of erg cm^{-2} s^{-1} Hz^{-1} or W m^{-2} Hz^{-1}. Although these units are the same as those of flux density (since J_ν is obtained by integrating over solid angle), the quantities are not the same.[4]

In the case of fully isotropic radiation, the integral in Equation (3.6) is $4\pi I_\nu$ and $J_\nu = I_\nu$.

Surface Brightness, B_ν

Whereas flux density is the measurement of the amount of radiation coming from a compact source, for an *extended* source (i.e., one that can be spatially resolved by the detector), the *surface brightness B_ν* is more useful. This is the flux density per unit solid angle, so the surface brightness is really an estimate of the specific intensity. It is generally quoted in observationally convenient units, often using square arcseconds as the solid angle, e.g., Jy arcsec^{-2}. There are 4.255×10^{10} square arcseconds per steradian.

---

[3] For example, the brightness of the infrared sky (a specific intensity) is often given in MJy sr^{-1}.
[4] Note the absence of the cosine term.

Luminosity, L

The luminosity of an astrophysical object is simply its power output in electro-magnetic radiation, measured in erg s^{-1} or W. Because this power is radiated outwards, a theoretical spherical shell around the object will capture all of its luminosity. If the flux is measured somewhere on that shell, it will be

$$F = \frac{L}{4\pi d} \qquad (3.7)$$

where d is the radius of the shell. If we set the radius of the shell to be the distance between the Earth and the source in question, then F will be the flux we measure on Earth. If, therefore, we measure the flux F of a source and know its distance from Earth d, we can calculate its luminosity—a fundamental astrophysical parameter—as

$$L = 4\pi \, d^2 F. \qquad (3.8)$$

The luminosity of stars and galaxies is generally very large, so astrophysical luminosities are generally given in units of Solar luminosity $L_\odot$:[5]

$$L_\odot = 3.845 \times 10^{33} \text{ erg s}^{-1} = 3.845 \times 10^{26} \text{ W}. \qquad (3.9)$$

Fluence, F

The *fluence* (often denoted F, but not the same as flux density) is defined as the energy passing through a unit area due to a flux of radiation or particles over time. Because it describes the total energy passing through, rather than a steady rate of energy flow, it is generally used to characterize transient events, which do not have a steady-state brightness, and are bounded by a start and finish, such as fast radio bursts (FRB) or gamma radio bursts (GRB).

Gamma-ray fluences are measured in units such as erg cm^{-2} or J m^{-2}, i.e., integrated over the energy spectrum. However, radio fluences are often quoted in units such as Jy s, which implies that it is measured per unit frequency.

3.1.2 The Magnitude System

In order to study the physics of astronomical objects, the most useful quantity to measure is the flux or flux density. In principle, we can simply give the flux density of the source in SI units. In practice, flux measurements or estimates are often given on the *magnitude scale*. This historical scale was developed by analogy with human vision, and persists because its logarithmic form allows a wide range of fluxes to be handled in a simple notation.

---

[5] Although luminosity strictly refers to power output by radiation, solar luminosity is sometimes used as a convenient unit of power in other astrophysical contexts, e.g., the "mechanical luminosity" of outflowing gas.

The Development of the Magnitude Scale

Hipparchus[6] is thought to have been the first to assign some sort of number to describe the brightness of the stars. In his star catalog, first-magnitude stars were the brightest and sixth-magnitude stars were just visible to his eye. The human visual response to brightness is more logarithmic than linear, so a difference of one magnitude roughly corresponded to a constant factor in physical brightness.

William Herschel put Hipparchus' scale on a more precise footing. He compared two stars, one of first magnitude and one of fifth magnitude, and determined that he had to decrease the aperture of the telescope with the brighter star to 1/100 that of the telescope with the dimmer star to make the two stars appear the same in brightness.

This result was converted to the modern magnitude scale in 1856 by Pogson.[7] If 5 magnitudes of difference correspond to a factor of 100 in physical brightness, then 1 magnitude corresponds to a factor of $100^{(1/5)} = 2.512$.

Apparent Magnitude

The apparent magnitude scale is now defined by the equation:

$$\Delta m = m_1 - m_2 = -2.5 \log \frac{F_1}{F_2} \tag{3.10}$$

so that two sources with fluxes F_1 and F_2 will have magnitudes differing by Δm, with the brighter source having the lower magnitude. This is a relative scale, in which two sources are measured in the same way and then compared. To create an absolute scale, a zero-point can be defined—a flux that corresponds to a magnitude of zero. If the zero-point is F_0, then a source of flux F has magnitude

$$m = -2.5 \log \frac{F}{F_0}. \tag{3.11}$$

This measurement scale does not take account of instrumental effects, and so it relies on standard methods of measurement to allow the necessary comparisons. These standard methods constitute *magnitude systems*, and their definition includes the equipment used to measure sources and the standards used.

The original reference star, defined to have magnitude zero in all bands, is Vega (α Lyrae), a main-sequence star of spectral type A0, and one of the brightest in the sky.

Unfiltered Magnitudes

The average human eye is most sensitive to wavelengths around 550 nm (5500 Å green light), with decreasing sensitivity above and below this. Measuring magnitudes corresponding to the response of the eye gives the *apparent visual magnitude*, m_V.

---

[6] Hipparchus of Nicaea, second century BCE, Greek astronomer.
[7] Norman Pogson (1829–1891), British astronomer who produced the Madras catalog of 11015 stars.

Photographic plates are most sensitive to the blue end of the optical spectrum but can also detect wavelengths not seen by the eye. Thus, the *photographic magnitude*, m_{pg} is usually different from the visual magnitude. Using a yellow filter can make the photographic plate response come closer to the human eye's. This is called the *photovisual magnitude*, m_{PV}.

Magnitude Measurements
An instrument for measuring the flux of a compact source will ideally measure:

$$\int_{\Delta\nu} F_\nu d\nu \qquad (3.12)$$

i.e., the flux density integrated over the observing bandwidth. This assumes that the observing bandwidth is defined by a perfect filter, with a transmission of 1 inside the band and 0 outside. In reality, filters are not perfect, and can be defined by their transmission profile, which is a function of frequency (or wavelength). In addition, the instrument sensitivity will vary as a function of frequency. So the flux measurement actually made will be:

$$\int_\nu F_\nu \, T(\nu) \, s(\nu) d\nu \qquad (3.13)$$

where $T(\nu)$ and $s(\nu)$ are the filter transmission and instrument sensitivity functions, respectively.

For magnitude measurements to be comparable to each other, they must be made with similar filters, and this has led to the definition of standard filters. These were originally specific colored glasses, but modern filters are mainly interference filters, in which multiple layers of material are stacked to build up interference patterns that allow transmission only within a defined bandpass.

The most common magnitude system is the *UBVRI* system, originally created in 1953 by Johnson[8] and Morgan[9] as the *UBV* system. It was then extended into redder colors by the addition of R and I. The letters stand for "ultraviolet," "blue," "visual," "red," and "infrared" respectively. The infrared referred to here is invisible to the human eye, but well within the frequency range of CCD sensors, which are the most common instruments used. Y and Z bands have also been added for CCDs with extended red sensitivity; Z has no upper limit in wavelength, using the CCDs long-wave cutoff instead.

The Johnson–Cousins[10]–Glass[11] system extends the filters used farther into the infrared region, into a regime requiring specialized IR detectors rather than CCDs. The added filters are labeled *J, H, K, L,* and *M*. This became the standard system of photometry and is the system actually used when speaking of the *UBV* system.

---

[8] Harold Lester Johnson (1921–1980), American astronomer.
[9] William Wilson Morgan (1906–1994), American astronomer. Also worked with P. Keenan to develop the MK system of spectroscopic stellar classification.
[10] Alan William James Cousins (1903–2001), FRAS, South African astronomer.
[11] Ian Glass (1939–) is an infrared astronomer and scientific historian from South African.

Table 3.1. Optical and Near-IR Filter Systems

ν (THz)	λ (nm)	T (K)	Name	Bandwidth (nm)
UBV (RI)				
817	367	7870	Ultraviolet (U)	66
688	436	6630	Blue (B)	94
550	545	5300	Visible (V)	88
470	638	4530	Red (R)	138
376	797	3630	Infrared (I)	149
uvby				
857	350	8260	Ultraviolet	30
730	411	7030	Visible	19
642	467	6190	Blue	18
548	547	5280	Yellow	23
Johnson–Cousins–Glass				
246	1220	2370	Infrared (J)	213
184	1630	1770	Infrared (H)	307
137	2190	1320	Infrared (K)	390
87	3450	838	Infrared (L)	472
63	4750	608	Infrared (M)	460

Note. 1 nm = 10 Å.

Variants have emerged for many of the filters. Kron–Cousins R and I are slightly different to the Johnson R and I. The near-IR K filter is only useful at very high sites (such as Maunakea)—lower sites tend to use K' or K_s filters, which are shifted to slightly shorter wavelengths. A narrower version of L, narrow-band L or nbL, is more commonly used than the normal L filter, and so on.

Filters—and hence magnitude systems—can be designed for specific purposes. The Gunn *griz* system (adopted by the Sloan Digital Sky Survey, SDSS, among others) avoids a lot of atmospheric optical emission, and so is particularly useful for observing faint galaxies, while the relatively narrow-bandwidth Strömgren[12] *uvbyβ* system is optimized for stellar classification (Tables 3.1 and F.4). Other systems include Vilnius, Geneva, and Walraven.

AB Magnitude: Since the magnitude scale can work off any zero-point, it is possible to define the zero-point, and hence the rest of the scale, in physical units. The AB magnitude scale (m_{AB}) does just that:

$$m_{AB} = -2.5 \log F_\nu - 48.60. \tag{3.14}$$

This is equivalent to a zero-point flux density of around 3631 Jy.

---

[12] Bengt Strömgren (1908–1987), Danish astronomer. Director of Yerkes and McDonald observatories, 1951–1957. Took Einstein's office in Princeton, 1957.

3.1.3 Color Indices

Since the use of standardized filters allows us to estimate the magnitude of an object at a specific wavelength—or color of light—the magnitudes may be compared to give a rough measurement of the color of the object itself; a coarse measurement of its spectrum.

The first *color index* used was defined as the difference between photographic and visual magnitudes (approximately, blue and yellow): $CI = m_{pg} - m_V$.

With the development of the *UBV* system, color indices could be defined as $U - B$ and $B - V$. As more colors were added to the system, more color indices were defined. Because of the negative logarithmic scale used for magnitudes, a redder star will have higher (more positive) $U - B$ and $B - V$, while a bluer star will have lower (more negative) indices. This is shown in Figures 3.2 and 3.3.

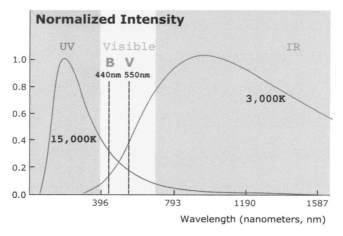

Figure 3.2. Comparison of spectrum of two stars with normalized intensity through B and V filters. Image credit: Adapted by M. Guidry.

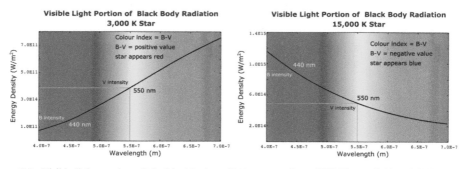

Figure 3.3. Visible light portion of the blackbody radiation curve for a 3000 K star (left) and 15,000 K star (right). Image credit: Adapted by M. Guidry.

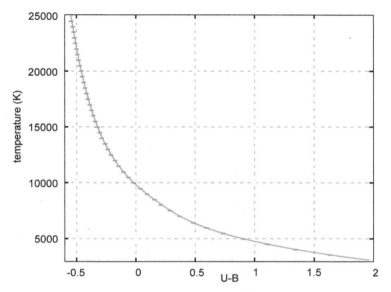

Figure 3.4. Plot of $U - B$ against temperature, normalized so that $U - B = 0$ at Vega's temperature of $T = 10,000$ K. Image credit: University of Nebraska-Lincoln https://astro.unl.edu/naap/blackbody/color-index.html.

In Figure 3.3(left), the lower B intensity value and higher V intensity value will give a positive $B - V$. Likewise the value of $B - V$ for Figure 3.3(right) will be negative. Because the star Vega is used as the zero-magnitude reference star, it follows that its colors will also be zero, so a 10,000 K blackbody will have $U - B = B - V = 0$ (see Figure 3.4).

Color indices, although very approximate measurements of a spectrum, have great utility in sorting and ordering objects, e.g., through color–magnitude and color–color diagrams. The use of magnitudes and colors outside their traditional optical domain is a demonstration of this.

Bolometric Magnitude
If we were to measure the flux over all frequencies, we would get the *bolometric magnitude*, m_{bol}.[13] The bolometric magnitude can be derived from the visual magnitude by the *bolometric correction*, *BC*:

$$m_{bol} = m_V - BC. \tag{3.15}$$

The bolometric correction for class F5 stars is, by definition, zero.[14] The reference flux for bolometric magnitude is greater than that for visual magnitude, therefore the

---

[13] There is a device called a *bolometer*, but it is sensitive mostly to infrared and submillimeter-waves, not the entire spectrum.
[14] Almost solar-like stars. The Sun is a G2 class star.

bolometric flux is always greater than the visual flux. The more an object's spectrum differs from that of the Sun, the greater the bolometric correction for the object.

3.1.4 Absolute Magnitude

The apparent magnitude m, does not take account of the distance to the object. In order to compare objects in terms of actual energy production, we also have the *absolute magnitude* scale, M. The absolute magnitude of a star is equal to its apparent magnitude when the star is ten parsecs[15] from Earth.

The difference between apparent and absolute magnitude is the *distance modulus*, $m - M$. This is related to the distance of the object from the observer. Since the flux density depends on the source distance from the surface at which it is detected, the flux of the source of interest, compared to the flux of the same source 10 pc away, is:

$$\frac{F(D)}{F(10 \text{ pc})} = \left(\frac{10 \text{ pc}}{D}\right)^2. \tag{3.16}$$

Using this in (3.11) to define the *distance modulus*, $m - M$ we find:

$$m - M = -2.5 \log \frac{F(D)}{F(10 \text{ pc})} = -2.5 \log \left(\frac{10 \text{ pc}}{D}\right)^2. \tag{3.17}$$

If D is not measured in parsecs, a conversion to parsecs must be included in this calculation.

Absolute magnitude is generally denoted with capital letters. However, in the *UBV* magnitude system, *U*, *B*, and *V* are apparent magnitudes. Thus, we use M_U, M_B, and M_V to denote absolute magnitudes in this system.

3.2 Measurements of Other Messengers

3.2.1 Cosmic Rays and Extensive Air Showers

The term *cosmic ray* is usually used to denote relativistic charged particles originating outside the solar system. The energy range of these cosmic rays is mostly from 100 MeV to 10 GeV (IceCube Masterclass 2020), but much higher energies have been recorded (Chapter 7). The discovery of these particles is generally attributed to V. F. Hess who flew charged electroscopes in balloons high into the atmosphere; he found the rate of electroscope-discharge increased significantly above 2000 m altitude, due to the ionization of the air by cosmic rays. They are now observed both from the ground and from space—satellites are used to observe the lower energies, and ground-based detectors observe the higher energies (Section 1.3.5).

Cosmic-ray detection generally includes an estimate of the energy, measured in electron-volts (eV; more commonly MeV and GeV). Counting cosmic ray detections and estimating their energy allows for a direct estimate of the flux of cosmic rays in GeV cm^{-2} sr^{-1} s^{-1}. The flux can also be defined as a particle flux, similar to a flux

---

[15] One parsec is the distance to a star having a stellar parallax angle of 1 s of arc. 1 pc = 3.26 lt-yr.

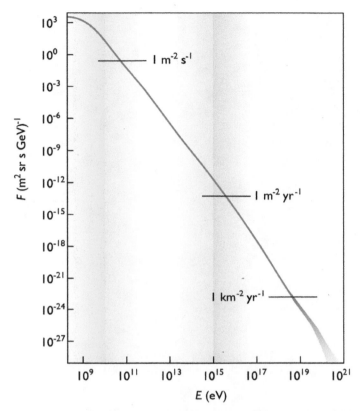

Figure 3.5. The flux of cosmic ray particles as a function of their energy. Image credit: Wikimedia (Sven Lafebre) CC BY-SA 3.0 https://commons.wikimedia.org/wiki/File:Cosmic_ray_flux_versus_particle_energy. svg.

density, but counting the number of particles passing through the surface rather than the energy. Figure 3.5 shows that this particle flux decreases sharply with increasing particle energy, so low-energy cosmic rays are much more common than high-energy ones. The flux for the lowest energies (yellow zone in Figure 3.5) are mainly attributed to solar cosmic rays, intermediate energies (blue) to galactic cosmic rays, and highest energies (purple) to extragalactic cosmic rays. The particles in most powerful man-made accelerator, LHC, can be accelerated up to energies of a few TeV.

Thus, for low-energy cosmic rays, the measuring surface does not need to be large, but the particles are absorbed by the atmosphere, so space-based instruments perform best. For high-energy cosmic rays, a much larger surface is required, which necessitates ground-based instrumentation, and the absorption of the cosmic ray by the atmosphere in an *air shower* is the detection method (Figure 3.6). This air shower is caused by the cosmic ray depositing its energy in the atmosphere, causing multiple new particles to be produced, which cascade downwards to the surface.

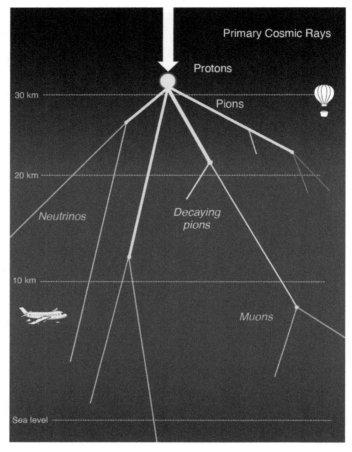

Figure 3.6. Primary cosmic ray interacting with particles in atmosphere and split into shower of secondary particles. Balloon experiments are used to detect primary rays.

Cosmic rays are charged particles, so their trajectory is influenced by magnetic fields as they travel through space. This is in contrast to electromagnetic radiation, which always travels in straight lines. Thus, photons arriving at an instrument from a specific direction may be traced back along that direction to their source. Cosmic rays, however, arriving from a specific direction, may be from a source lying in a very different direction. In order to use them as cosmic probes, therefore, different measurements must be used, such as the elemental abundance. About 90% of cosmic rays are protons, 9% are helium, and the remaining 1% include heavier elements (Chapter 7).

Cosmic rays could be also defined as the relativistic non-thermal plasma—a "local fog" of Ultra High Energy Cosmic Ray (UHECR) produced by nearby Galactic sources such as SNRs, Pulsars, Galactic center, etc. (Liu et al. 2016). Together with matter, radiation, and magnetic fields, cosmic rays could be seen as the "4th substance" of the visible universe.

3.2.2 Neutrinos

Neutrinos are chargeless elementary particles with a small mass (see Chapter 7 and Sections 1.3.7 and 7.7). they therefore travel in straight lines, like photons, and unlike cosmic rays. They are, in fact, even more direct messengers of cosmic physics than photons, because they have very small cross-sections for all interactions. For example, neutrino fluxes from supernova explosions escape immediately from the supernova, whereas many photons are trapped and scattered. The quantities measured for neutrinos include their energy (in eV), the direction from which they radiate toward us, and the type of neutrino.

There are three types, or flavors, of neutrinos: electron neutrinos, muon neutrinos, and tau neutrinos, which are related to other, charged, fundamental particles— electron, muon, and tau. If a neutrino interacts with a matter particle, its related charged-particle partner (e.g., electron for electron neutrino) is released. The flavors of neutrinos interchange between each other as they travel (IceCube Masterclass 2020).

In Earth's atmosphere, neutrinos are very abundant (they are second only to photons in abundance throughout the universe). Yet they are extraordinarily hard to detect, simply because they have such small interaction cross-sections with matter. Measurement of cosmic messengers is essentially a business of having the messengers interact with a piece of material that we have instrumented, so the only way to detect neutrinos is to have a very large amount of matter instrumented, so that the few interactions that take place are recorded.

The actual detection of a neutrino is by its secondary effects. The neutrino interacts with an atom, and other particles are produced, which are much more detectable than the neutrino itself. One of the most common ways to detect the secondary particles is by their Čerenkov radiation, as they travel greater than the speed of light in the medium in which they are immersed (often water or water–ice). Of course, such secondary particles are also produced by other high-energy events, such as cosmic rays, so discrimination between the very rare neutrino interactions and the much more common non-neutrino interactions is a central problem in neutrino astrophysics.

As with cosmic rays, the flux of neutrinos drops rapidly with neutrino energy (Figure 3.7). The highest-energy neutrinos are particularly important to our understanding of the Universe, but are also hard to detect, simply because they are much rarer.

3.2.3 Gravitational Waves

Gravitational waves are a result of the acceleration of massive bodies (see Chapter 8). The properties of the detected waves therefore trace the masses and movements of these bodies. For a gravitational wave to be detectable, the accelerating bodies must have at least stellar-scale mass, and must be compact objects, such as neutron stars or black holes. While the gravitational waves spread out across space, becoming harder to detect, they are not otherwise affected by their passage through the universe, so, like neutrinos, they give a clear picture of the processes that produce them.

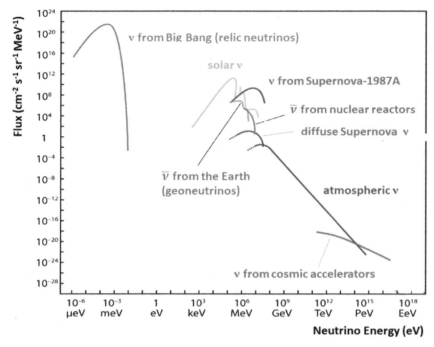

Figure 3.7. An explanatory image of measured and expected flux values of neutrinos at various energy levels. Image credit: IceCube Collaboration/Christian Spiering masterclass.icecube.wisc.edu/en/learn/detecting-neutrinos.

The properties measured for these waves are the same as that of any wave: frequency and amplitude. The wave is measured by its effect on test masses, separated by long distances. The relative positions of the masses are monitored by laser interferometry. There are many other processes than can affect the position of test masses, and these must all be rejected in order to see the gravitational wave signal. The combination of data from multiple detectors is one method to do this, and also allows for a more precise estimate of the direction of travel of the wave, and hence where it comes from on the sky.

References

IceCube Masterclass, 2020, Detecting Neutrinos, https://masterclass.icecube.wisc.edu/en/learn/detecting-neutrinos

IceCube Masterclass, 2020, Measuring Cosmic Rays, https://masterclass.icecube.wisc.edu/en/icetop/measuring-cosmic-rays

Liu, R.-Y., Taylor, A. M., Wang, X.-Y., & Aharonian, F. A. 2016, PhRvD, 94, 043008

Chapter 4

The Transfer of Electromagnetic Radiation Through Space

To understand the data from our telescopes, we must understand all the physical mechanisms that interact with the radiation on its way from its original source to our detectors. The purpose of this chapter is to provide the reader with an overview of the physical processes involved in the interaction of matter and radiation transport. These processes may be part of the physics of the astronomical object or they may be the result of interactions with the interstellar or intergalactic medium.

Again, we do not go into any depth in discussion of the physics and derivation of the associated formulae. We begin with fundamental ideas, and then the interactions between the atoms (or molecules) and their environments that may cause changes in the emitted or absorbed spectral photon frequencies or wavelengths. Then we consider the scattering of photons by their interactions with matter encountered while on their way to our detectors.

4.1 The Doppler Effect and Redshifts

When there is a relative velocity between a source and an observer of waves, there can be a noticeable shift in the frequency of the waves. The frequency and the speed of the emitted wave are not altered—they are still defined by the physics of the emission mechanism. The observer, however, measures a different frequency or wavelength.

In 1842, Doppler[1] published a monograph discussing the "...colored light of the binary refracted stars..." in which he describes a change in wavelength of waves,

---

[1] Johann Christian Andreas Doppler (1803–1853), Austrian mathematician and physicist.

doi:10.1088/2514-3433/ac087ech4

caused by relative motion between source and observer. For astronomical observations (in his monograph), the relative motion would be caused by the Earth's motion around the Sun and so his discussion involved Bradley's[2] discovery of stellar aberration. This shift in wavelength has since been known as the Doppler effect, and it was experimentally confirmed in sound waves by Ballot[3] and in light by Fizeau.[4] For light waves, the observed frequency is given by:

$$\nu = \nu_0 \left(1 - \frac{u}{c} \right) \tag{4.1}$$

where ν is the observed frequency, ν_0 is the emitted frequency and u is the relative velocity between the source and observer.

When the source and observer are approaching, the frequency is increased and so the velocity is negative. This is called *blueshift*. When the source and observer are receding, the frequency is decreased and so the velocity is positive. This is called *redshift*. It is purely a matter of convention that redshifts and blueshifts correspond to positive and negative velocities. In terms of wavelength, this formula becomes:

$$\lambda = \frac{\lambda_0 c}{c - u}. \tag{4.2}$$

These formulae are only first order approximations to the effect, valid only in the case where $\nu << c$ and the distance between the source and observer is not small. A more correct form of the equations is created using the special theory of relativity.

Relativistic Doppler Effect: The relativistic form of the Doppler equations include the effect of time dilation and therefore these equations contain the required Lorenz symmetry of the special theory. For motion along the line of sight:

$$\nu = \sqrt{\frac{1 - u/c}{1 + u/c}} \, \nu_0 \tag{4.3}$$

where again, approaching velocities are negative. The corresponding wavelengths are given by:

$$\lambda = \sqrt{\frac{1 + u/c}{1 - u/c}} \, \lambda_0. \tag{4.4}$$

---

[2] James Bradley (1693–1762), discovered the aberration of starlight in 1728, caused by the orbital motion of the Earth.
[3] Christoph Ballot (1817–1890), Dutch physicist.
[4] Hippolyte Fizeau (1819–1896), French physicist; in France, the Doppler effect is called "effet Doppler–Fizeau."

For motion in an arbitrary direction:

$$\nu = \frac{\nu_0}{\gamma \left[1 + \dfrac{u \cos \theta_0}{c} \right]} \qquad (4.5)$$

or:

$$\lambda = \lambda_0 \gamma \left[1 + \frac{u \cos \theta_0}{c} \right] \qquad (4.6)$$

where:

$$\gamma = \frac{1}{\sqrt{1 - u^2/c^2}} \qquad (4.7)$$

and u is the velocity with which the source is receding from the observer at an angle θ_0 relative to the line of sight at the time the light was emitted. If $\theta_0 = 90°$ we obtain the *transverse* Doppler effect:

$$\nu = \frac{\nu_0}{\gamma} \quad \text{or} \quad \lambda = \lambda_0 \gamma. \qquad (4.8)$$

If we measure the velocity to line-of-sight angle in the source's reference frame and the time of observation (when the light is received by the observer), θ_s, the expression becomes:

$$\nu = \gamma \left[1 - \frac{u \cos \theta_s}{c} \right] \nu \qquad (4.9)$$

and θ_0 and θ_s are related to each other by *relativistic aberration*:

$$\cos \theta_0 = \frac{\cos \theta_s - \dfrac{v}{c}}{1 - \dfrac{v}{c} \cos \theta_s}. \qquad (4.10)$$

If $\cos \theta_s = 0$, relativistic aberration gives $\cos \theta_0 = -v/c$ and substituting this into (2.77) gives the other form of the transverse Doppler effect:

$$\nu = \nu_0 \gamma \quad \text{or} \quad \lambda = \frac{\lambda_0}{\gamma} \qquad (4.11)$$

which is generally considered more useful.

The Redshift Index, z: The redshift index, z is defined as:

$$z + 1 = \frac{\lambda_0}{\lambda} = \frac{\nu}{\nu_0} = \sqrt{\frac{1 + u/c}{1 - u/c}}. \qquad (4.12)$$

4.1.1 The Doppler Effect in Astronomical Observations

The Doppler effect can be used in observations to determine the radial component of the relative velocity of the source and the observer. From the equations above:

$$v = c\left[\frac{\nu_0^2 - \nu^2}{\nu_0^2 + \nu}\right] \qquad (4.13)$$

with the direction of the velocity radial component assigned to be positive for redshift and negative for blueshift.

Not only is this a reliable way to measure velocities of remote objects, it can be very precise. If the source frequency is well-known, the observed frequency can be measured to any desired precision, giving similarly precise velocity measurements. The Doppler effect is therefore the foundation of our understanding of the kinematics of the universe. It is used to measure the velocities of stars in the Galaxy, the reflex motions of stars due to orbiting exoplanets, and the recession of galaxies in the universe.

For example, the observation of Doppler shift in spectra can be used to measure the rate at which a star is approaching or receding from the Sun. The two extremes here are the stars BD–15° 4041 approaching at 308 km s^{-1} and Wolf 1106, receding at 260 km s^{-1}. It is also used to determine the rate of expansion rate of planetary nebulae (PNe) or SNRs. If the rate of expansion and distance to such objects are known, we can calculate the age of those objects. This is where observations of extragalactic SNRs and PNe have an advantage, because the distances to extragalactic nebulae are known to better relative accuracy than those to galactic nebulae.

4.2 Radiative Transfer and Optical Depth

As electromagnetic radiation passes through a medium (such as an interstellar cloud or the inside of a star), some of it may be absorbed, so some of the energy flux of the radiation is transferred to the medium. The medium may also emit radiation, which will add energy flux to the radiation. Radiation may also be scattered in the medium; this results in a change of direction of a given photon, and may also transfer energy between medium and radiation.

Radiative transfer is a very general treatment of the transport of energy by electromagnetic radiation through a medium with which it interacts. The general theory does not consider the microphysics of the interaction—the mechanisms by which radiation may be absorbed or emitted by the medium. It merely treats the bulk behavior of the medium and the radiation energy flux through it.

Absorption: As the electromagnetic radiation passes through a medium, some fraction of it will be absorbed. The fraction of the intensity that is absorbed per unit distance at a specific frequency is the *opacity* κ_ν, so that, as the radiation penetrates the medium by a small distance ds, the intensity change is

$$dI_\nu = -\kappa_\nu I_\nu ds. \qquad (4.14)$$

This differential equation can be solved to give the intensity on the path through the medium:

$$I_\nu(s) = I_\nu(0)e^{-\tau_\nu} \tag{4.15}$$

where $I_\nu(0)$ is the intensity at the start of the path through the medium (usually the intensity coming into a cloud), and τ_ν is the *optical depth* or *optical thickness*:

$$\tau_\nu = \int_S \kappa_\nu ds \tag{4.16}$$

or $\kappa_\nu S$ in the case of a homogeneous medium. While the opacity is a property of the medium itself, the optical depth is a property of the object, such as an interstellar cloud. An object with low optical depth (*optically thin*) is relatively transparent to radiation, merely attenuating it; background sources can be seen through it. An object with high optical depth (*optically thick*) is opaque.

The opacity can also be related to the microscopic properties of the medium. If the medium is thought of as containing n absorbers per unit volume, each with an absorption cross-section σ_ν, $\kappa_\nu = n\sigma_\nu$. This can be integrated over the path length: $\tau_\nu = N\sigma_\nu$, where N is the column density of the absorbers.

Emission: Whereas the amount of radiation depends on the incoming radiation intensity, the amount of radiation emitted—the added intensity—depends only on the properties of the medium itself:

$$dI_\nu = j_\nu ds \tag{4.17}$$

where j_ν is the emissivity of the medium, which depends on its physical properties.

Emission and Absorption: In general, emission and absorption will occur in the same medium, with a radiative transfer equation:

$$\frac{dI_\nu}{ds} = -\kappa_\nu I_\nu + j_\nu . \tag{4.18}$$

Local Thermodynamic Equilibrium: In local thermodynamic equilibrium (LTE), the constituents all have the same temperature (i.e., distribution over energy states), with the exception of the radiation field (or photon gas), which will generally have a different temperature. In LTE, Kirchoff's law, relating emissivity to absorptivity, applies:

$$\frac{j_\nu}{\kappa_\nu} = B_\nu(T) \tag{4.19}$$

where $B_\nu(T)$ is the Planck blackbody function at temperature T. If temperature is constant throughout the medium, the radiation intensity follows:

$$I_\nu(s) = I_\nu(0)e^{-\tau_\nu(s)} + B_\nu(T)(1 - e^{-\tau_\nu(s)}). \tag{4.20}$$

The most common use of this equation is for a medium emitting radiation, such as a gas cloud. In this case, the absorption term is not important, and if the cloud is optically thin ($\tau \ll 1$),

$$I_\nu \approx \tau_\nu B_\nu(T) \tag{4.21}$$

but if the cloud is optically thick ($\tau \gg 1$),

$$I_\nu \approx B_\nu(T).$$ (4.22)

4.3 Scattering Processes

4.3.1 Rayleigh Scattering

From the evidence of interstellar extinction, we see that blue light is scattered much more strongly by the interstellar medium than red light. The scattering approximately follows the characteristic λ^{-4} law of Rayleigh scattering. Light generally Rayleigh-scatters off very small obstructions, much smaller than the wavelength. These may be atoms, in which the electron bound to the nucleus acts as a harmonic oscillator (essentially, a mass on a spring); or they may be small objects—generally treated as spheres.

Rayleigh Scattering by a Bound Electron: Low energy photons may be absorbed and very quickly re-emitted by an electron bound to an atom (or molecule) without the atom changing to an excited state. We can see this interaction in a classical understanding as the interaction of the photon with the electron only. In such a view, the electron is considered attached to the atom by a spring and the electron acts as a driven, simple harmonic oscillator. The electron re-radiates the energy, scattering the direction of energy flow.

If ν_0 is the characteristic or resonant frequency of the system and ν is the frequency of the photon, the Rayleigh cross-section is then:

$$\sigma_R = \frac{8\pi}{c}\left[\frac{e^2}{m_e c^2}\right]^2 \frac{\nu^4}{(\nu^2 - \nu_0^2)^2}.$$ (4.23)

As can be seen, the scattering is very frequency dependent ($\sigma_R \propto \nu^4 \propto \lambda^{-4}$).

Rayleigh Scattering by a Small Sphere: If the radiation of wavelength λ is incident on a small dielectric sphere (an ideal dust particle) of radius a, where $a < \sim 0.05\lambda$, the scattering cross-section is given by:

$$\sigma_s = \frac{128\pi^5 a^6}{3\lambda^4}\left[\frac{n^2 - 1}{n^2 + 2}\right]^2$$ (4.24)

where n is the refractive index of the sphere.

The scattering cross-section is simply the effective area of the scattering particle, and can be compared to its geometric area to give a scattering efficiency

$$Q_s = \frac{\sigma_s}{\pi a^2}.$$ (4.25)

If the incident radiation is linearly polarized and the angle between the plane of polarization and propagation, and the observer, is θ, the observed intensity of the scattered radiation is given by:

$$I = \frac{16\pi^4 a^6}{D^2 \lambda^4}\left[\frac{n^2 - 1}{n^2 + 2}\right]^2 \sin^2\theta$$ (4.26)

where D is the distance to the observer and the incident radiation has unit intensity. If the incident radiation is polarized, the scattered radiation is also polarized. If the incident radiation is not polarized, the scattered radiation can be partially polarized.

4.3.2 Thomson Scattering

Thomson scattering[5] is the absorption of energy from an electromagnetic wave by a charged particle and the subsequent re-emission of radiation by the acceleration of that particle due to the absorption of the wave's energy. The direction of the re-emitted radiation is not related to the incident wave's direction of propagation and hence, the radiation is scattered. Thomson scattering is an important phenomenon in plasma physics.

When the EM wave strikes a charged particle (e.g., an electron), the varying electric and magnetic fields accelerate the particle, which then causes the particle to emit photons. If the motion of the electron is not relativistic, the acceleration of the electron is due mainly to the electric field component of the EM wave. The particle's motion will then be directed by the oscillating electric field and the re-emitted radiation will have a dipole nature, polarized in the direction of the electron's motion. The frequency of the scattered radiation is equal to the frequency of the incident radiation.

The scattering process can be analyzed by considering a *scattering volume*; a volume dV of charged particles. If the light incident on it has flux density per unit wavelength F_λ (i.e., power per unit area per unit wavelength, over all angles), it will scatter light into a solid angle element $d\Omega$ at an angle ϕ to the incoming light. The power of the scattered light per unit wavelength, $dP_\lambda = \varepsilon dV \, d\Omega \, d\lambda$, where ε is the *emission coefficient*. Integrating over solid angle, the scattered light power per unit wavelength per unit scattering volume per unit wavelength is

$$dP_\lambda = \frac{8}{3}\pi^2 F_\lambda \sigma_T \, n \tag{4.27}$$

where n is the volume density of charged particles in the scattering volume, σ_T is the *Thomson cross-section*

$$\sigma_T = \frac{8}{3}\pi\left(\frac{q^2}{4\pi\epsilon_0 mc^2}\right)^2 \tag{4.28}$$

and q and m are the charge and mass of a charged particle.

The emission coefficient is comprised of two components, ε_t which emits light polarized perpendicular to the scattering plane (*tangential*), and ε_r which emits light polarized in the scattering plane (*radial*). ε_r has a dependency on $\cos^2 \phi$, leading to a degree of polarization in the scattered light.

---

[5] Named for Sir Joseph John Thomson (1856–1940), discoverer of the electron. He was a student of Lord Rayleigh and won the 1906 Nobel Prize in physics.

4.3.3 Compton Scattering

Compton scattering[6] (or the *Compton effect*) is the decrease in energy (decrease in frequency or increase in wavelength) of an X-ray or gamma-ray photon when it interacts with a free, charged particle (usually an electron). The change in the photon's frequency is called the *Compton shift*.

Thomson scattering (Section 4.3.2) cannot explain the shift in energy (frequency) of the photon as it requires the scattered radiation to have the same frequency as the incident radiation. As Compton demonstrated, the only way to explain the shift is to assume the particle nature of light (photons), where the energy carried by these particles is proportional to their equivalent wave frequency. Compton scattering is most noticeable with photons in the 0.5–3.5 MeV ($\nu = 1.2$ to 8.5×10^{20} Hz) range.

Using conservation of momentum and energy applied to a collision problem between a high-energy photon and a free electron, Compton derived this relationship between the before and after frequencies of the photon:

$$\frac{1}{\nu_a} \frac{1}{\nu_b} = \frac{h}{m_e} c^2 (1 - \cos \theta) \tag{4.29}$$

where ν_a and ν_b are the after and before interaction frequency of the photon and θ is the angle through which the photon is scattered. If the electron was at rest, it recoils with a velocity proportional to the energy absorbed from the photon.

4.4 Interstellar Extinction and Reddening

While the scattering processes described above are microscopic, they give rise to macroscopic effects as light travels to us through the interstellar and intergalactic media. There are two linked large-scale effects: the light of stars is dimmed as it is absorbed on its way to us or scattered out of our line of sight to the star (*extinction*); and the extinction is more effective at shorter wavelengths, so that the star appears redder than it really is (*reddening*).

Both of these effects are mainly due to interstellar dust, and the measurement of extinction and reddening can be used to characterize the dust. By removing these effects, the brightness and color of stars and galaxies can also be more accurately estimated (Figure 4.1).

Interstellar dust has a broad distribution of grain sizes, and so interstellar scattering will be a combination of Rayleigh scattering (for grains smaller than the wavelength) and Mie scattering (for grains comparable to or larger than the wavelength). While the Rayleigh scattering cross-section has a very strong wavelength dependence (λ^{-4}), the Mie scattering cross-section does not have a monotonic

---

[6] Named for its discoverer, Arthur Holly Compton (1892–1962), Professor of physics at Washington University, St. Louis, MO. The effect was discovered in 1923 and it confirmed the particle properties of light, suggested by Planck and Einstein. Compton won the 1927 Nobel Prize in physics for this discovery.

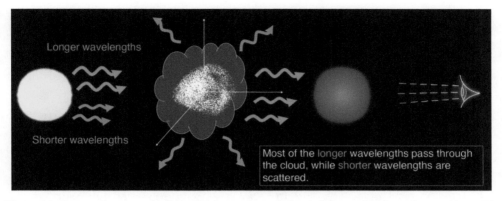

Figure 4.1. Dust grains along the line of sight scatter and absorb light coming from distant objects. We therefore see these objects as dimmer and redder than they really are. These effects are known as extinction and interstellar reddening respectively.

wavelength dependence. So the overall wavelength dependence is still strongly wavelength-dependent, and additional wavelength-specific features are found, which can be attributed to the solid-state physics and chemistry of the dust grains, e.g., the 2175Å bump, the diffuse interstellar bands, the 3.1 μm water ice feature, the 10 and 18 μm silicate features.

Extinction: Extinction is inherently hard to measure, since we can only measure the dimmed stellar brightness. In order to estimate the amount of the dimming, the undimmed brightness must be estimated first. This can be done by, for example, spectroscopically classifying the star to estimate its inherent brightness.

The extinction is a property of a line of sight from the Earth to an astrophysical object, and arises from all the matter along the line of sight that can reduce the light intensity. The extinction measures the ratio of the observed intensity I to the inherent intensity I_0. Because this is a brightness ratio, it can be expressed in the UV–optical–near-IR domain as a change in magnitude:

$$A = m - m_0 = -2.5 \log \frac{I}{I_0} \qquad (4.30)$$

(since $I_0 > I$, $m > m_0$ and $A > 0$).

The extinction can be defined at any wave band. For historical reasons, the extinction in the V wave band, A_V, is most often used.

Reddening: The measurement of reddening, similar to that of extinction, requires some knowledge of the basic spectral shape of the astrophysical source observed. The reddening is often given as a *color excess*, E, which is analogous to the extinction A, but for color rather than magnitude. Just as an extinction can be defined in any wave band, a color excess can be defined for any color index, but the most commonly used color excess is $E(B - V)$:

$$E(B - V) = (B - V) - (B - V)_0. \qquad (4.31)$$

More dust along the line of sight causes an increase in both extinction and reddening, and they are generally related by the ratio of total to selective extinction R. For the V band,

$$R_V = \frac{A_V}{E(B - V)}.$$ (4.32)

R_V is generally found to be 3.2, although slightly different values have been measured along some lines of sight.

Although reddening is generally estimated from broadband measurements, it can be estimated from spectroscopic measurements. The ratio of brightnesses of two spectral lines can be compared to their theoretical ratio, and the discrepancy used to calculate the reddening. The Hα/Hβ ratio is a commonly used line pair for this technique.

If the reddening and hence extinction can be estimated toward many stars, the results can be used to map the distribution of matter in the interstellar medium. This technique has been used in the near-IR, where the vast majority of stars have the same color index, and so any variation in the observed color index can be attributed to reddening and hence the presence of interstellar dust.

Chapter 5

The Earth's Atmosphere

Because most astronomical observations are made with telescopes within the Earth's atmosphere, it is wise for astronomers to know at least the basics of the atmosphere's structure and character. The atmosphere is after all, the reason we need space-based observatories.

5.1 The Large-scale Structure of the Atmosphere

The Earth's atmosphere consists of mainly nitrogen (78.1%) and oxygen (20.9%), with small amounts of argon (0.9%), carbon dioxide (variable, approximately 0.035%), water vapor, and other gases such as neon at 0.0018% (Figure 5.1). The atmosphere protects life by absorbing harmful ultraviolet, X-ray, and gamma-ray solar radiation and preventing extreme temperature changes between night and day. The same mechanisms that protect life necessitate observations in much of the spectrum to be made from space-based observatories.

The atmosphere has no abrupt upper boundary. Rather, it gradually thins out with increasing altitude. There is no definitive boundary between the atmosphere and space; different characteristics of space start at different altitudes: at about 19.5 km the blood of an unprotected human (i.e., one exposed to ambient pressure) begins to boil. For NASA, anyone who travels above 80 km is designated an astronaut. Atmospheric effects become noticeable during spacecraft re-entry at about 120 km. One frequently used altitude for a somewhat arbitrary boundary between atmosphere and space is 100 km, which is also known as the Kármán line.[1] 99.9% of the atmosphere lies below this boundary (Figure 5.2).

---

[1] Named for Theodore von Kármán (1918–1963), Hungarian–American engineer and physicist, researcher in aeronautics and astronautics. This is the altitude at which, because of the low density of the atmosphere, an aircraft would need to travel faster than orbital speed to generate enough lift for flight.

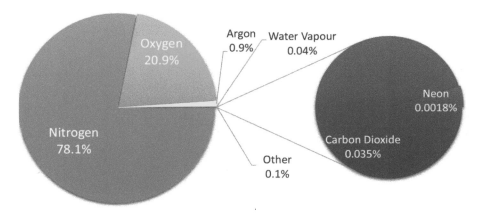

Figure 5.1. Percentage composition of the Earth's atmosphere.

5.1.1 Main Layers of the Atmosphere

Because the chemical composition varies with altitude, and these chemicals have greater and lesser interactions with sunlight, the temperature of the Earth's atmosphere both increases and decreases with altitude. It is the turning points in the gradual temperature changes that mark the layer boundaries of Earth's atmosphere. Table 5.1 shows the altitude of the layer boundaries. These altitudes have seasonal and latitudinal (between equator and poles) variations. The average temperature at the surface is 14° C.

The *troposphere* (from the Greek word tropos, meaning turning) is the lowest layer of Earth's atmosphere and the one in which clouds and most weather phenomena occur. It contains approximately 80% of the total air mass, and nearly all the atmosphere's water vapor and particulate matter. Because of temperature changes, this layer reaches 16–18 km over tropical regions, decreasing to less than 10 km over the poles. The troposphere is divided into six zonal flow regions, called cells. These cells are responsible for the atmospheric circulation producing the prevailing winds. The change of temperature with height is larger than in other layers.

The *tropopause* marks the boundary between the troposphere and the *stratosphere*.

Air temperature within the stratosphere increases with altitude, reaching 280 K at the top. The *stratosphere* (from French stratosphère—"sphere of layers"; Ancient Greek: sphaĩra—"sphere") is a layer of intense interactions among radiative, dynamic, and chemical processes. The horizontal mixing of gases proceeds much more rapidly than vertical mixing. A good portion of weather balloons reach and operate at this layer.

The temperature increases rapidly near the top of the stratosphere primarily because of the stratospheric *ozone layer* (or ozone shield), which absorbs solar ultraviolet radiation (this absorption is a major reason for the need for UV astronomy to be done from space). The ozone layer also holds sulfides, which are used in the alignment of some of the active optics systems of large optical telescopes. The discovery of ozone layer in 1913 is attributed to the French physicists

Figure 5.2. Layers of Earth's atmosphere. Image credit: NOAA (User:Mysid).

Table 5.1. The Major Layers of the Earth's Atmosphere (also see Figure 5.2)

Layer	Boundary	Altitudes (km)	Comments
Troposphere		0–7/20	300 K–235 K
	Tropopause	7–17/20	Variation in altitude is mostly latitudinal.
Stratosphere		7/17–50	235 K–280 K
	Stratopause	50	
Mesosphere		50–80/85	80 K–170 K
	Mesopause	80–85	
Thermosphere		80/85–690	170 K–2800 K. Temperature is very sensitive to solar radiation.
	Exobase	690	
Exosphere		>640	Because of the extremely low density, temperature has little meaning. Low-mass atoms (H, He) are escaping into space.

Charles Fabry (1867–1945) and Henri Buisson (1873–1944). Commercial airplanes fly in the lower stratosphere as the less-turbulent layer provides a smoother ride.

Within the *mesosphere* (Greek mesos, "middle"), temperature decreases with increasing altitude. Temperatures in the upper mesosphere fall to ~180 K (~−90° C; the coldest temperatures in Earth's atmosphere) and vary with latitude and season. Most meteors burn up in the mesosphere.

In the *thermosphere* (based on the Greek word for heat) the temperature reaches its extreme due to absorption of highly energetic solar radiation by the residual oxygen. The residual atmospheric gases form strata according to their molecular mass. Solar ultraviolet radiation causes ionization creating the ionosphere. Also, a large number of human-made satellites orbit Earth within the thermosphere as well as the famous space-shuttle. Finally, the aurora (a.k.a. the Southern and Northern Lights) primarily occur in this layer.

In the *exosphere* (Ancient Greek: éxō—"outside, external, beyond"), the atmosphere blends into space. The few particles of gas are mainly hydrogen and helium in the process of escaping the Earth's gravity. Their high speed can lead to measures of high temperature, but only because the density is so low, leading to very long mean free paths. These particles are in "ballistic" trajectories and under the influence of gravity, which means that some of them could escape into space.

5.1.2 Other Atmospheric Layers

There are a few other atmospheric layers that are of concern to astronomers. These are listed in Table 5.2.

The Ionosphere

In the thermosphere, the density is so low that free electrons can exist for extended periods of time before recombining with ions. This layer of high ion and

Table 5.2. Other Layers of Earth's Atmosphere of Concern to Astronomers

Ionosphere	The layer containing ions produced by solar ionization, roughly 50 to 400 km altitude. Blocking radio.
Airglow	These layers cause emission lines in optical spectra taken with ground-based telescopes.
Ozone layer	Approximately 10–50 km. Blocking UV.
Magnetosphere	The region where the Earth's magnetic field interacts with the solar wind from the Sun. The upper most atmospheric layer.
Van Allen belts	Regions where protons (lower altitude) and electrons (high altitude) from the solar wind are trapped by the Earth's magnetic field.

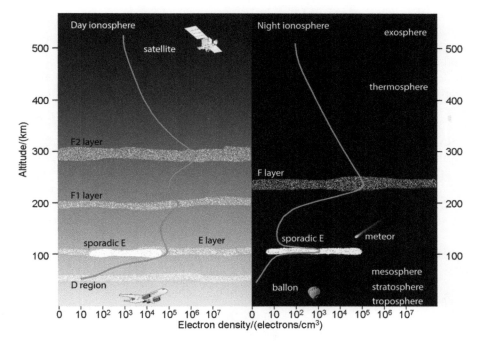

Figure 5.3. Layers of Earth's ionosphere. High frequency waves are reflected by the ionosphere at a height of between 100 and 400 km, depending on the radio frequency and electron density of the ionosphere. Image credit: Aleksandar Zorkić.

free-electron count is called the *ionosphere*, and consists of the D, E, E_S, and F layers[2] (Figure 5.3).

The D layer is the lowest layer at altitudes between 50 and 90 km, where hydrogen Lyman-series alpha radiation (121.5 nm) from the Sun is ionizing nitric oxide (NO).

---

[2] The E layer is also known as the Kennelly–Heaviside layer. This layer was predicted to exist, independently and simultaneously by Arthur Kennelly (1861–1939) and Oliver Heaviside (1850–1925). Sir Edward Appleton (1892–1965) won the 1947 Nobel physics prize for demonstrating the existence of the ionosphere in 1924. The F layer is also known as the Appleton layer.

An active Sun with 50 or more sunspots provides hard X-rays that also ionize the N_2 and O_2. Due to the higher gas density in the D layer, the recombination of ions and electrons is high so the net ionization effect is low. Thus, high-frequency radio waves are not reflected by the D layer—but rather absorbed—particularly at 10 MHz and below, with decreasing absorption with higher frequency. The absorption is least at night and greatest at midday. The ionization in this layer is greatly reduced after sunset, but galactic cosmic rays cause it to persist.

The E layer lies between 90 and 120 km in altitude. The ionization of this layer is due to soft X-ray (1–10 nm) and far-ultraviolet (100–280 nm) radiation from the Sun, ionizing oxygen (O_2). This layer can only reflect (not absorb) radio waves having frequencies less than about 10 MHz. It partially absorbs frequencies above 10 MHz. The E layer also tends to disappear (but not completely) at night because of the loss of the solar ionizing radiation.

The E_S layer is also called the sporadic E-layer and is consists of small clouds of intense ionization causing reflection of frequencies between 25 and 225 MHz. Sporadic-E events may last from a few minutes to several hours. Research has found multiple causes of sporadic-E events, most of which are still not understood. Major E_S events occur most often during the summer months with only minor events during the winter.

The F layer lies 120 to 400 km in altitude. Extreme ultraviolet (10–100 nm) solar radiation ionizes atomic oxygen (O). The F region is the most important part of the ionosphere in terms of high-frequency radio observations. During the day, sunlight causes the layer to divide in two, labeled the F_1 and F_2 layers. These combine into one layer at night. The F layer causes most skywave propagation of radio waves, and are thickest and most reflective of radio during the daytime.

Because the ionosphere reflects both cosmic radio waves and terrestrial radio waves, it can be troublesome to radio astronomy and its effects could color astronomical radio data if completely ignored. Fortunately, most radio signals of interest lie outside of those mostly affected by the ionosphere.

The Airglow Layers (Atmospheric Emissions)

The *airglow* layers are found at (roughly) 80, 193, and 290 km altitude (Figure 5.4). They create weak, emission-line optical and near-infrared (OH^- molecules have strong infrared emissions) spectra. This airglow limits the sensitivity of all ground-based telescopes. Airglow emissions are reasonably uniform across the atmosphere, but to an observer on the ground appear brightest at ten degrees above the horizon. This is related to atmospheric mass. Below ten degrees, atmospheric extinction reduces the airglow. In optical astronomy, the brightness of airglow is described in terms of apparent magnitude per square arcsecond.

There are a number of causes for airglow. One is the recombination of ions created during the daylight hours, with free electrons. Another is the recombination of ions created by cosmic rays, with free electrons. A third source is chemical processes occurring in the atmosphere. Radiation from the Sun disassociates N_2 and O_2 molecules. Then individual N and O atoms can combine to form NO with the

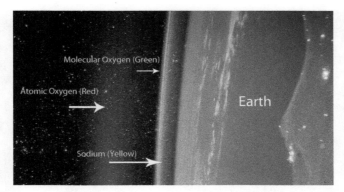

Figure 5.4. Excited oxygen at higher altitude creates a layer of faint red airglow. Sodium excitation forms the yellow layer at 80 km up. Airglow is brightest during daylight hours but invisible against the sunlight sky. Image credit: NASA with annotations by Alex Rivest.

emission of a photon. The NO molecule then contributes to blocking the infrared and submillimeter bands from reaching the ground. There are other chemical reactions in the upper atmosphere that also produce airglow.

The unit of measure for airglow is the Rayleigh,[3] denoted R. The Rayleigh is a column emission rate of 10^{10} photons per square meter per second. This is purely an emission rate and there is no accounting for absorption or scattering. Airglow on a typical night is about 250 R, but the aurora may reach 1 MR.

Near-infrared Airglow: Airglow is not particularly important at visible wavelengths, but the hydroxyl (OH^-) airglow emission bands dominate the sky background in the near-infrared, shortwards of about 2.3 μm (longwards of 2.3 μ, the sky background is dominated by water vapor emission and the thermal emission of the atmosphere itself). The OH airglow constitutes a major limit on the ability of ground-based telescopes to see faint objects at the J and H near-IR bands. The airglow fluctuates on spatial and temporal scales, so that it cannot simply be removed from the data. Significant effort has gone into devising ways to defeat the effects of OH airglow on near-IR astronomy.

The Ozone Layer

The *ozone* layer is a shell of protective gases around the Earth and inside the stratosphere. It allows for the existence of land-based life on Earth because it blocks the Sun's ultraviolet radiation. While this is a good thing for astronomers in general, it also means that ultraviolet astronomy must be done from space-based platforms. It lies from 20 to 50 km in altitude. Solar ultraviolet light breaks O_2 into separate

---

[3] Named in 1956 after Robert Strutt (1875–1947), 4th Baron Rayleigh, son of the Lord Rayleigh of Rayleigh scattering.

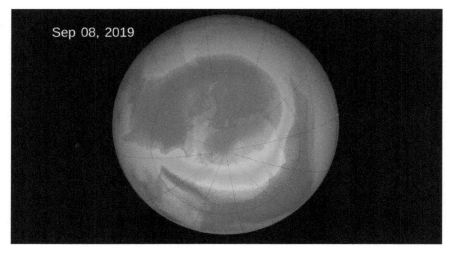

Figure 5.5. Scientists from NASA track the ozone layer throughout the year and determine when the hole reaches its annual maximum extent. In 2019, unusually strong weather patterns caused warm temperatures in the upper atmosphere above the South Pole region of Antarctic, which resulted in a small ozone hole. Image credit: NASA's Goddard Space Flight Center.

atoms that combine with unbroken O_2 to form O_3 (ozone). This molecule is then broken by ultraviolet light back into O and O_2, which recombines to form O_3 again, and the cycle continues. Most UVA (315–400 nm) (see Appendix Table F.3) reaches the Earth's surface. The ozone layer is significantly more effective at blocking UVB (290 nm), but some UVB still reaches the surface. UVC (100–280 nm) is entirely removed by ozone at the altitude of about 35 km. The reduction in ozone over polar regions—particularly Antarctica—due to chemical reactions driven by pollution is being reversed, and Figure 5.5 shows the "ozone hole" from 2019, which is the smallest on record since its discovery.

5.1.3 The Magnetosphere

Some refer to the *magnetosphere* as the outermost layer of the atmosphere (Figure 5.6). The magnetosphere, however, is characterized not by atmospheric molecules, but with the interaction of the Earth's magnetic field with the stream of solar wind particles. Solar wind particles can have the same effect on our telescopes' detectors as cosmic rays. The orbits of space-based telescopes are generally within the magnetosphere which can then aid in the avoidance of solar wind particles. However, the high density protons and electrons of the Van Allen radiation belts, which are part of the magnetosphere, must be avoided. The near space environment must then be considered when determining the orbits of our space observatories. Any detector events caused by solar wind particles must be removed from the data along with the cosmic rays, which are essentially unstoppable.

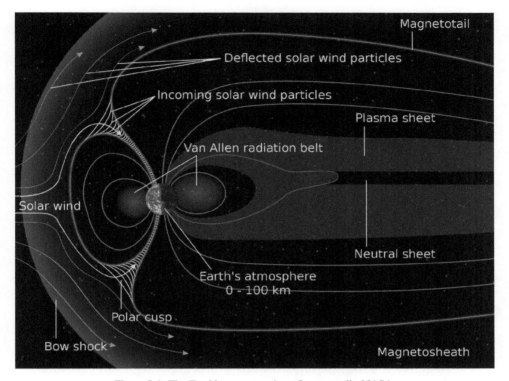

Figure 5.6. The Earth's magnetosphere. Image credit: NASA.

5.1.4 Physical Characteristics of the Atmosphere

Pressure: The air pressure of the atmosphere at a given altitude depends on the weight of air above that altitude. As the altitude increases, the pressure decreases, and the density of the air drops, so that there is less mass in each unit of height of the atmosphere. The overall effect is of an exponential drop in air pressure with altitude, to a good approximation. At sea level, atmospheric pressure averages about 100 kPa. The scale height of the exponential drop (the altitude at which the pressure drops by a factor of e) is 6000 to 8500 m, depending on temperature. About 50% of the total mass of the air lies below 5 km altitude, so the air pressure around 5 km (the altitude of the highest ground-based telescopes) is about half that at sea level.

Density and Mass: The density of air at sea level is about 1.255 kg m^{-3}, decreasing with altitude as the pressure decreases. The total mass of the atmosphere is about 5.1×10^{18} kg, a tiny fraction of Earth's total mass (about one millionth).

Atmospheric Refraction: The refractive index of air varies with pressure, temperature, and the amount of water vapor, but is around 1.000 3 (sometimes written as $(n - 1) \times 10^6 \approx 300$) in the UV–visible–infrared domain. The apparent position of an astrophysical source in the sky will be shifted by refraction by an amount of order an arcminute over most of the sky, but the amount of refraction grows rapidly as the source gets closer to the horizon and the path length through the atmosphere

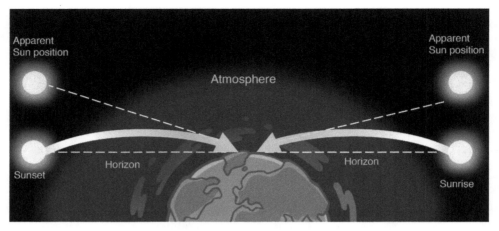

Figure 5.7. Atmospheric refraction.

increases, getting to about a degree as the source rises or sets. This is generally compensated for by telescope control systems. Atmospheric refraction also shifts the apparent position of radio sources (Figure 5.7).

As extended light sources, like the Sun or Moon, approach the horizon, their images are distorted as different parts of the image are refracted differently. This *differential refraction*, in which different subsets of the light from a source suffer different amounts of refraction, is a larger-scale form of the problem of astronomical *seeing*.

5.2 Turbulence and Seeing

5.2.1 Astronomical Seeing

Point-spread Function and Angular Resolution: When the light from a point-like source passes through an optical system, it emerges in a spatial distribution defined by the *point-spread function* or PSF, which is a function of angle on the sky. The PSF effectively defines the angular resolution of the system—if two point sources are so close to each other that their PSFs largely overlap, it becomes hard to separate the PSFs accurately, and the sources are essentially indistinguishable. The sources are not resolved.

In the case of an electromagnetic-wave telescope, and in absence of an atmosphere, the PSF has the form of an *Airy disk*,[4] whose size is inversely proportional to the diameter of the telescope aperture. Thus, the inherent resolution of a telescope increases with aperture.

The Atmospheric PSF and Seeing: The transmission of light through the atmosphere also constitutes an optical system, and a star observed at the bottom of the atmosphere follows the PSF generated by the atmosphere. This PSF follows an approximately Gaussian distribution, and depends not on the telescope, but on

---

[4] Named after George Biddell Airy (1801–1892).

the atmosphere above it. The size of this atmospheric PSF, usually given as the full width at half maximum (FWHM), is the quantity usually referred to as seeing. A seeing of 1.0″ is a reasonable one for standard astronomical locations. Some of the best seeing is found at high altitude on small islands such as Maunakea (Hawaiian Islands) or La Palma (Canary Islands). These locations can have a seeing FWHM as low as ~0.4″.

The atmospheric PSF is the result of turbulent motion in the atmospheric column above the telescope. The turbulent mixing of air masses at different temperatures and densities causes different subsets of the light coming from a source to be subject to different refractive paths through the atmosphere, before finally arriving at the telescope aperture. They therefore appear to come from different directions, and the addition of all those smaller images creates the larger Gaussian seeing PSF. Figure 5.10 shows the blurring effect of seeing.

Astronomical seeing is an inherently spatial and temporal effect, caused by the averaging over time and space of multiple subsets of the light from the observed object, as they follow their different paths through the atmosphere. Sometimes the temporal nature of seeing can be seen, as when stars twinkle in the sky.

Seeing has been among the biggest challenges for Earth-based telescopes, despite the advancement of larger instruments that theoretically provide millisecond resolutions. There can be a factor of 100 between theoretical and actual resolution because of these atmospheric seeing limitations. Overcoming the effects of seeing, by selecting better locations and by technological development, is a major preoccupation of ground-based astronomers.

5.2.2 The Turbulent Structure of the Atmosphere

Atmospheric turbulence, caused mainly by convection currents, mixes atmospheric layers with different temperature and chemical composition, which in turn causes small (in magnitude and location) variations in the atmospheric refraction (Figure 5.8). The air's index of refraction is a function of density which is far more sensitive to changes in temperature than in pressure. This is true both in the free atmosphere and in the air near or in the telescope.

Fluid flows can be characterized by the dimensionless *Reynolds number*:

$$Re = \frac{VL}{\eta} \tag{5.1}$$

where V is the fluid velocity, L is a length scale and η is the viscosity of the fluid. The Reynolds number therefore compares the inertial forces in the fluid to the viscous forces. In the case of a high Reynolds number, parcels of fluid in motion will tend to keep moving, leading to a flow state with different parcels moving in different directions with different speeds, or *turbulent flow*; with a low Reynolds number, parcels of fluid will get dragged along with the general fluid flow, leading to a flow state with smoothly-varying velocities, or *laminar flow*.

A layer where characteristics differ between these two types of flow is known as a boundary layer—like the bounding surface in a pipe's interior. The injection of a

Figure 5.8. Starry Night painting by Vincent van Gogh. Turbulence among the stars? Image credit: The Starry Night, Vincent Van Gogh/Wikipedia.

high-velocity fluid stream into a low-velocity fluid, such as hot gases released by a flame in the air, causes a similar effect (Figure 5.9). This difference in velocity induces fluid friction, which is a factor that allows turbulent flow to form. The viscosity of the fluid, which tends to prevent turbulence, counteracts this effect.

Viscosity is very small for air: $\nu = 1.5 \times 10^{-5} \, \mathrm{m^2 \, s^{-1}}$. Thus, for typical wind speeds and length scales of meters to kilometers, $Re > 10^6$ and the air is moving turbulently.

Lower Altitude Effects: The majority of turbulent airflow of the atmosphere occurs near the ground (0–100 m). The turbulence is caused by structures such as buildings composed of varying material and densities that radiate heat differently, grassy fields, hills, and bodies of water, resulting in convection currents as the Sun heats the ground during the day and that heat is radiated at night.

This "near ground seeing" can even be generated by the observing equipment and environment itself. A good example is "dome seeing," where air has been trapped in a telescope dome and heated during the day. At night, some of the heated air escapes and the rest cools down. The temperature contrasts can drive significant turbulence.

Mid Altitude Effects: These effects are found in the low troposphere (100–2000 m), and are determined by the topography around the observing site. Mountain ranges, large cities, and densely populated areas located upwind cause disturbances in the local atmosphere. Airflow as far as 100 km downwind of a mountain peak can give rise to turbulence. Thus, it is better to observe downwind of a flat surface such as a large body of water or desert, which will generate more laminar and less turbulent flow.

High Altitude Effects: In the high troposphere (6–12 km), "rivers" of fast moving air are known as jet streams. The large velocity gradients around them cause strong wind shear, adding extra turbulence. The location of jet streams shifts over time.

Figure 5.9. The plume goes from a laminar to a turbulent one from this candle flame. The Reynolds number could be used to anticipate where this transformation will happen. Image credit: Wikipedia: Gary Settles (CC BY-SA 3.0) https://en.wikipedia.org/wiki/Reynolds number.

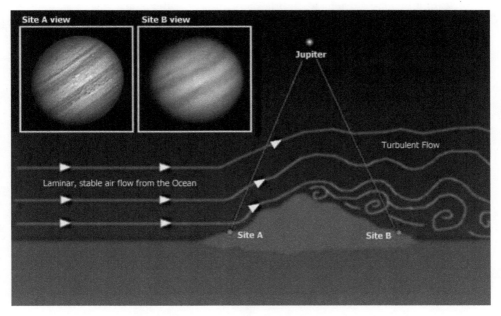

Figure 5.10. A diagram showing how mountains break up stable airflow into turbulence. Note the difference in the probable views from site A (facing into the prevailing winds off the ocean) and site B (located on the downwind side of the mountain peaks). Image credit: Damian Peach http://www.damianpeach.com/seeing1.htm.

5.2.3 Statistical Description of Turbulence

In different models of turbulent flows and optical aberrations, atmospheric turbulence has been well studied and described. These can be used to extract descriptive statistics on the degree of turbulence that can be achieved for both the turbulence itself and its impact on the quality of the image. Because optical turbulence and optical aberrations are distinct, and created by several different distortion sources, it is not possible to relate them in depth. Instead, the correlation is conveniently described as an average turbulence and average image aberration.

Three typical explanations of the conditions of astronomical seeing at an observatory exist and we describe them next.

Kolmogorov Theory of Turbulence: Eddy Cascade

Kolmogorov's[5] theory describes how energy is transferred from larger to smaller eddies. It also describes how much energy is contained by eddies of a given size and how much energy is dissipated by eddies of corresponding size.

Eddies of different sizes make up turbulence. An "eddy" eludes a precise definition, but it is conceived to be a turbulent motion, localized over a region of size l, that is at least moderately coherent over this region. Let us assume that energy is added to a system at the largest scales—"outer scale" L_0. Then energy cascades from larger to smaller scales (turbulent eddies "break down" into smaller and smaller structures). Size scales where this takes place are called "inertial ranges." Finally, the eddy size reaches the "inner scale," l_0, so small that it is subject to dissipation from viscosity. L_0 ranges from tens to hundreds of meters while l_0 is a few mm.

For any inertial range the energy spectrum shape is $k^{-5/3}$ (Figure 5.11; k is wave number). When turbulence occurs in a given atmospheric layer with a temperature gradient, it mixes air of different temperatures at the same altitude and produces various temperature fluctuations. Hence, the spectrum also describes the expected variation of temperature in turbulent air.

Let u be velocity, outer scale L, then:

$$\text{Energy} \sim u^2 \qquad (5.2)$$

energy dissipation rate:

$$E \sim \frac{u^2}{t} \sim \frac{u^3}{L}. \qquad (5.3)$$

Hence:

$$u^2 \sim (EL)^{2/3}. \qquad (5.4)$$

Taking its Fourier transform gives the power spectrum $L \sim k^{-1}$

$$f(k)dk \sim u^2 \sim k^{-2/3} \qquad (5.5)$$

---

[5] Andrey Nikolaevič Kolmogorov (1903–1987); his theory was first published in 1941 ("K41 theory") and later papers were in 1962.

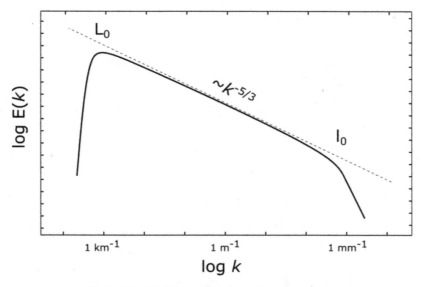

Figure 5.11. Depiction of the observed energy cascade.

or

$$f(k) \sim k^{-5/3}. \tag{5.6}$$

To understand turbulence in the context of seeing, we convert the above thermal turbulence spectrum into a spatial context. In other words, we examine how the atmosphere is going to transform wavefronts. The wavefront perturbations are brought about by variations in the refractive index of the atmosphere in the path of the light. These refractive index variations lead directly to "phase fluctuations" (dominant) and "amplitude fluctuations" (second-order effect that is negligible).

For monochromatic plane waves arriving from a distant point source with wavevector k, we have:

$$\psi_0(r, t) = Ae^{i(\phi_0 + 2\pi \nu t + kr)}. \tag{5.7}$$

The turbulent layer scatters some light, perturbs the phase of the wave and causes fractional amplitude change with effect:

$$\psi_p(r) = (\chi_a(r)e^{i\phi_a(r)})\psi_0(r). \tag{5.8}$$

Tatarskii Theory

The Kolmogorov model developed by Tatarskii (1961) offers a definition of the significance of the wavefront disturbances induced by the atmosphere. A number of experimental measurements (Buscher et al. 1995) support this model and it is commonly used in astronomical imaging simulations. The model suggests that changes in the refractive index of the atmosphere cause the wavefront perturbations.

These refractive index variations directly contribute to the phase fluctuations explained by $\phi_a(r)$, but any amplitude perturbations are only created as a second-order effect as the disrupted wavefronts spread to the telescope from the upsetting atmospheric layer. Instantaneous imaging efficiency is dominated by phase fluctuations $\phi_a(r)$ in all appropriate models of the Earth's atmosphere at optical and infrared wavelengths.

The phase fluctuations in the model of Tatarskii[6] are mostly believed to have a Gaussian, random distribution with following second-order structure function for simplicity:

$$D_{\phi_a}(\rho) = \langle |\phi_a(r) - \phi_a(r + \rho)|^2 \rangle_r \qquad (5.9)$$

where $\langle ... \rangle$ represents average over ensemble and $D_{\phi_a}(\rho)$ is the variation caused atmospherically between the phase at two sections of the wavefront isolated in the aperture plane by a distance ρ. Tatarskii (1961) structure function can be expressed by a single parameter r_0:

$$D_{\phi_a}(\rho) = 6.88 \left(\frac{|\rho|}{r_0} \right)^{5/3}. \qquad (5.10)$$

r_0 also refers to the aperture diameter for which the "variance" σ^2 of the wavefront phase approaches to unity (David 1965); averaged over the aperture:

$$\sigma^2 = 1.029\,9 \left(\frac{d}{r_0} \right)^{5/3}. \qquad (5.11)$$

Description of Turbulence Above Telescopes
r_0 **and** t_0: Equation (5.11) above reflects a widely used description of r_0, a parameter often used in astronomical observatories to characterize atmospheric conditions. r_0 can be found from a measured C_N^2 (see below) quantity as follows:

$$r_0 = \left(16.7\lambda^{-2}(\cos \gamma)^{-1} \int_0^\infty dh C_N^2(h) \right)^{-3/5} \qquad (5.12)$$

where turbulence strength C_N^2 changes by height h above the telescope, and γ is the angular distance of the astronomical origin from the zenith. Parameters r_0 and t_0 can easily define the astronomical observing circumstances at an observatory. The "Fried parameter"[7] is also referred to as the seeing parameter r_0. Typical measured r_0 values are 20 to 40 cm for the infrared band (900 nm wavelength) at good observing sites. The resolution of long-exposure images is computed mainly by diffraction and the size of the Airy configuration for telescopes with diameters greater than r_0, and is thus inversely proportional to the diameter of the telescope.

---

[6] Valerian Ilich Tatarskii (1929–2020).
[7] David L. Fried (1933–).

For telescopes having diameter higher than r_0, the resolution of the image is mainly determined by the atmosphere and is irrespective of the diameter of the telescope, staying constant at the value given by the telescope with a diameter equal to r_0. The length scale above which the turbulence becomes meaningful also relates to r_0, and t_0 refers to the timescale above which the turbulence variations become important. In a system of adaptive optics (AO), r_0 specifies the width of the actuators required, and t_0 establishes the correction speed required to compensate for the consequences of the atmospheric turbulence.

The parameters r_0 and t_0 change with the wavelength used during astronomical imaging, causing large telescopes to be used for significantly higher-resolution imaging at longer wavelengths (see example on Figure 5.12).

Refractive Index Structure Constant C_N^2: The statistical definition of the average atmospheric turbulence starts with the *temperature structure coefficient* c_T^2, which is a variation described by the average squared distance defined in three dimensions. It is evaluated as:

$$C_T^2 = \frac{[T(x) - T(x + r)]^2}{r^{2/3}} \tag{5.13}$$

where x can be any point in the turbulence, and r is the three-dimensional interval between x and any other point having similar temperature. Greater energy turbulence

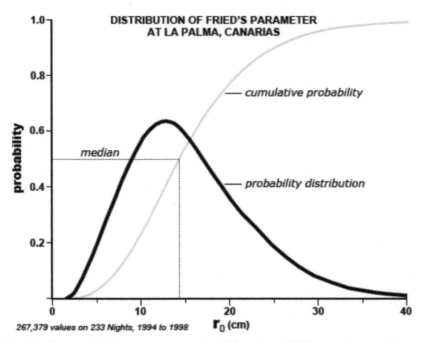

Figure 5.12. Distribution of Fried's parameter r_0 at the William Herschel Telescope site in the Canary Islands. Image credit: Reproduced from MacEvoy (2013) ©2013 Bruce MacEvoy, CC BY-ND 3.0. https://www.handprint.com/ASTRO/seeing1.html.

certainly splits "iso-thermal cells" into small parts, lowering the average distance between points with the same temperature: the average C_T^2 gets smaller as turbulence becomes more intense.

Similarly, a more precise explanation of an observatory's astronomical seeing is given by generating a turbulence strength profile often called a *refractive index structure constant*, denoted by c_N^2, as a function of altitude. It describes the average variations of refractive indices in the turbulent layer between different points. It is essentially the temperature structure coefficient determined by the average temperature and pressure of air. It is given as:

$$C_N^2 = C_T^2 \left[7.9 \times 10^{-5} \frac{P}{T^2} \right]^2 \tag{5.14}$$

where T and P are temperature (in Kelvins) and atmospheric pressure (in millibars).

If the volume of turbulence cells reduces, the refractive index gets lower, but exponentially decreasing as pressure decreases and temperature rises. In general, C_N^2 profiles are carried out when determining the type of adaptive optics for a specific telescope, or when deciding whether or not a specific site would be a suitable place to establish a new astronomical observatory. Usually, for calculating the C_N^2 profile, multiple methods are used and then compared simultaneously. Variations of C_N^2 with altitude h is shown in Figure 5.13.

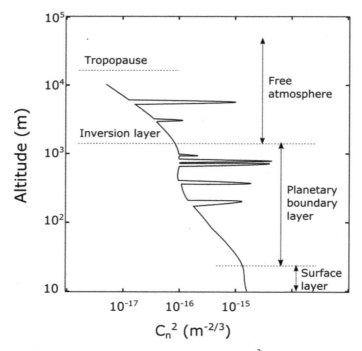

Figure 5.13. Variations of turbulence strength C_N^2 with altitude.

Seeing Disk as Full Width at Half Maximum: Finally, the "isoplanatic angle" θ, derived from Fried's parameter r_0 (Section 5.2.3), is the mean angular spacing of an isoplanatic region along a ground-based observer's optical path:

$$\theta = 0.6\left(\frac{r_0}{h}\right) \qquad (5.15)$$

where h is the altitude of layer of turbulence. This "isoplanatic angle" is the angle in the sky during which the turbulence outcomes are correlated (Figure 5.14).

5.2.4 Tackling Astronomical Seeing

The turbulent structure of the atmosphere manifests in several different seeing phenomena. These arise from the fact that the telescope generally sees an astrophysical source through several turbulent cells at once.

1. In a process known as *twinkling* or *scintillation* (Figure 5.15), a star's brightness tends to fluctuate. As the turbulent cells refract the light into or out of the telescope beam, the intensity of light, made up of the light traveling through all the cells in the beam, will change.

2. Because each turbulent cell refracts the light coherently—that is, there is no major phase variation in the wavefront from one side of the cell to the other—each cell creates a diffraction-limited Airy disk at the focal plane of the telescope, the size of the disk being defined by the size of the turbulent cell. Each of these disks is a *speckle*, and the image of a star is simply a number of speckles all moving around on the focal plane. An instantaneous

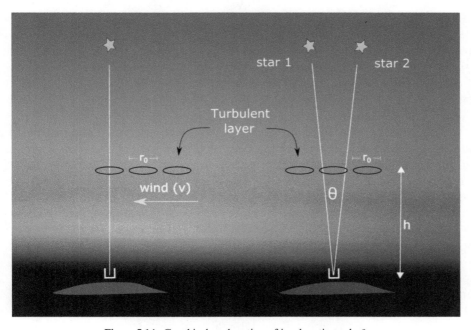

Figure 5.14. Graphical explanation of isoplanatic angle θ.

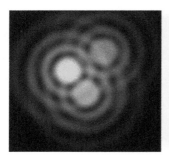 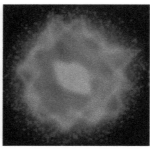 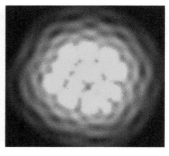

Figure 5.15. Description of various effects of astronomical seeing due to atmospheric turbulence. Scintillation (left), blurring (middle), and speckling (right). Image credit: Reproduced from MacEvoy (2013) ©2013 Bruce MacEvoy, CC BY-ND 3.0. https://www.handprint.com/ASTRO/seeing1.html

image would show several simultaneous Airy disks separated by dark boundaries (Figure 5.15).

3. If the image is exposed for longer than the time for the speckles to change, the overall effect will be the averaging of the speckle brightness together over time into a blurred, smooth, centrally-peaked distribution of light. This is the approximately Gaussian pattern that is the PSF of a telescope in the atmosphere (Figure 5.15).

Techniques to Reduce the Effect of Seeing
Adaptive Optics: The most common technique to alleviate the problem of seeing is *adaptive optics*, or AO. The effect of all the turbulent cells in the atmosphere above the telescope is to deform the wavefront, so that it arrives at different parts of the telescope mirror at different times. In AO, a deformable mirror is placed in the optical system, and its deformation is controlled to compensate for these time delays by changing the path lengths through the optical system. Because the mirror needs to be deformed quickly and accurately, it must be small and thin—these mirrors are placed into the telescope's optical train after the primary and secondary mirrors have concentrated the light into a small area.

The deformation of the AO mirror is carried out by actuators, and calculated by monitoring a *guide star*. This is a bright star that is not part of the science program, but can be assumed to be a point source; thus its image is a measurement of the PSF of the optical system at that time. The light from the star is diverted to a *wavefront sensor*, and the output of this sensor is used to calculate the deformation of the mirror required to correct the wavefront deformation. The simplest way to do this is *tip-tilt* correction, in which the mirror is only moved in angle. This very low-order correction is comparatively simple to implement, but can produce good results.

The use of AO absolutely depends on the selection of a suitable guide star. The more detailed the correction required, the brighter the star needs to be, and the harder it is to find. A solution to this comes with the use of a *laser guide star*, in which a laser is mounted on the telescope, and emits a beam of light into the atmosphere. The laser is either Rayleigh-scattered back to the telescope from the lower atmosphere, or it excites a patch of the sodium layer in the mesosphere. Both of these alternatives provide an artificial guide star.

Speckle Imaging: At any given time, the image of a star appears in a telescope as one or more speckles, each of which is much higher-resolution image than the time-averaged blurred image. By directly analyzing these speckles, higher-resolution information can be accessed.

If a star is bright enough that it can be imaged with an exposure time comparable to the timescale of the speckle movement, a series of short images can be recorded. Then each one can be shifted, so that the speckles on the images are lined up together, giving a sharper PSF. This technique is known as "shift-and-add."

Shift-and-add can be elaborated into *lucky imaging* that exploits the fact that some of the samples in a shift-and-add image set will have much higher resolution, simply because the random nature of atmospheric turbulence leads to some being blurred less than others. By selecting those "lucky" images for shift-and-add, a higher resolution can be obtained, at the cost of ignoring much of the data taken.

Speckle Interferometry is a more sophisticated technique in which the power spectrum of the speckles is Fourier-analyzed to infer the structural properties of the emitting source. For bright sources with simple structure, much higher resolution can be achieved.

Interferometry: While aperture synthesis interferometry has become a workhorse technique in radio astronomy to overcome the diffraction limit, it is a much less common technique in the optical and IR. Several interferometers have been built to carry it out, mainly using small telescopes (e.g., COAST in the UK, CHARA and NPOI in the USA). It has also been implemented on larger telescopes, where there is more than one telescope available (e.g., the European VLTI in Chile, the Keck Interferometer in Hawaii, and the Large Binocular Telescope Interferometer in Arizona).

Intensity interferometry is an even less common technique, using the time dependence of optical emission to constrain the diameters of stars.

5.3 Atmospheric Windows

The atmosphere is largely opaque to electromagnetic radiation. There are two major wavelength ranges, or "windows", where it is effectively transparent—the optical and radio windows. The short-wave cutoff of the optical window is around 300 nm in the near-UV; light with shorter wavelength than this is absorbed by the ozone layer; shorter wavelengths (100–230 nm) are absorbed by air molecules (nitrogen and oxygen gas), and even shorter wavelengths (<100 nm) are absorbed by oxygen and nitrogen atoms at high altitudes. The air remains an opaque absorber throughout the high-energy spectrum, until the very highest energies—very high-energy gamma-rays—which penetrate deeply enough into the atmosphere that their interaction can be seen from the ground as Cherenkov light.

The long-wavelength cutoff of the radio window is defined by the ionosphere that reflects electromagnetic radiation with frequency below its plasma frequency of about 15 MHz (Section 5.1.2).

The optical window starts to close around 1 μm wavelength, and the atmosphere does not become transparent again until the short-wave edge of the radio window around 15 mm. In between those windows, however, the opacity of the atmosphere varies greatly, and observations can often be carried out in between the optical and radio windows, in the infrared, terahertz, submillimeter-wave, and millimeter-wave regions. Radiation in these ranges is absorbed by atmospheric molecules, most notably oxygen, carbon dioxide, and water. Oxygen and carbon dioxide are ubiquitous, and so absorption by these species is fairly constant and very high. Water vapor, on the other hand, has a low scale height, and its abundance varies considerably over the Earth's surface, so wavelengths that are absorbed by water can often be observed from sites that are high, dry, or both. Within this range, the atmosphere can be characterized not so much as transparent or not, but by the altitude to which the radiation can penetrate before absorption. Figures 5.16 and 1.1 give a graphical representation of the penetration of the atmosphere by various frequencies.

As the optical window closes, the near-infrared bands, lying between water vapor absorptions, are still observable from mountain sites. As the wavelength increases, this gets harder and harder: ground-based observation is rare past 2.5 μm, and becomes impossible past about 25 μm.

Very similar effects are seen at the short-wave end of the radio window. The first water vapor absorption is found at around 22 GHz (14 mm), and then another window opens, which rapidly closes around 30 GHz (10 mm). The atmosphere then becomes somewhat transparent at a succession of windows between water vapor absorptions. These windows have wavelengths of about 7 mm, 3 mm, 1 mm, 0.85 mm, 0.45 mm, 0.35 mm, and 0.2 mm. As the wavelength shortens, the transparency of these windows

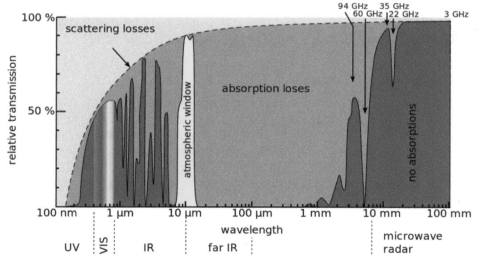

Figure 5.16. The Earth's atmosphere blocks most of the radiation of interest to astronomers. The line represents the altitude at which these frequencies are absorbed along with a brief description of the atoms or molecules responsible for the absorption. Image credit: Wikipedia: Cepheiden.

drops, until the 0.2 mm window (in the Terahertz spectrum) is only accessible from a very few sites in Chile and Antarctica.

Between about 25 μm and 200 μm, it is essentially impossible to observe from the ground, but airborne telescopes may be able to observe, depending on the altitude and the wavelength of interest.

5.3.1 Atmospheric Opacity

Up to the long-wavelength ionospheric cutoff (at which the atmosphere becomes reflective), atmospheric opacity is due to the ability of the atmospheric gases to absorb radiation at a particular wavelength. The process can be described using the theory of radiative transfer (Section 4.2). In general, the atmosphere both absorbs and emits at a given wavelength, and absorption and emission are linked.

The optical depth depends not only on the opacity of the atmosphere at that wavelength, but the path length through the atmosphere: if a source is observed toward the zenith, the path length through the atmosphere is simply the depth of the atmosphere; but if the source is observed below the zenith, the radiation takes a slanted path through the atmosphere, and the path is longer. We can define the *airmass* as the effective path length in units of the zenith path length: at zenith, airmass $A = 1$, and below the zenith $A > 1$.

If the atmosphere is treated as a horizontal slab, and the source is observed at a *zenith angle* z ($z = 90° - e$, where e is the source elevation), the airmass $A \approx \sec z = 1/\cos z$. The slab approximation breaks down close to the horizon, where $z \approx 90°$ and $\sec z$ increases without limit. In fact, the Earth's atmosphere is a (roughly) spherical shell, and so the path length cannot become infinite.

Atmospheric Absorption
At short wavelengths (up to ~2 μm), the blackbody emission function becomes very small, and emission from the atmosphere becomes negligible. The atmosphere is simply an absorber, and the brightness of the radiation received at the ground can be found from Equation (5.16) or (4.15):

$$I_\nu = I_\nu(0)e^{-A\tau_\nu} \tag{5.16}$$

where $I_\nu(0)$ is the brightness above the atmosphere, A is the airmass, and τ_ν is the optical depth of the atmosphere at zenith. If $\tau_\nu \gg 1$, the atmosphere is effectively opaque, and no radiation penetrates to the surface.

Extinction Correction: If τ_ν is small but not negligible, the radiation will penetrate to the surface but be attenuated. For $\tau_\nu \ll 1$,

$$I_\nu \approx (1 - A\tau_\nu)I_\nu(0) \tag{5.17}$$

so a plot of I_ν against A for a given source (generally a star) will be a straight line with a gradient of $-\tau_\nu$. Although the lowest measured A will be 1, the line can be extrapolated to airmass zero to give an estimate of $I_\nu(0)$, the brightness outside the atmosphere, thus correcting for atmospheric extinction.

Atmospheric Emission

Through most of the infrared to millimeter-wave spectrum, atmospheric emission is important—indeed, at submillimeter wavelengths, sky emission is much brighter than the astrophysical radiation being observed. The radiation brightness at the surface will follow Equation (4.20):

$$I_\nu = I_\nu(0)e^{-A\tau_\nu} + B_\nu(T)(1 - e^{-A\tau_\nu}) \tag{5.18}$$

where T is the atmospheric temperature, and local thermodynamic equilibrium is assumed. The atmospheric temperature, of course, varies over the path length, so this is a further approximation.

Because the optical depth τ_ν appears in both emission and absorption terms, it can be derived from measurements of the atmospheric emission, and then applied to correct I_ν to $I_\nu(0)$, the radiation brightness above the atmosphere.

Skydips: A common calibration technique in the millimeter-wave to terahertz spectrum is the *skydip*. This is similar to the extinction correction, in that the brightness is measured as a function of airmass, but it is the overall sky emission that is measured, not the brightness of a source attenuated by the atmosphere. The attenuated astrophysical contribution (i.e., the absorption term) can generally be neglected, so:

$$I_\nu \approx B_\nu(T)(1 - e^{-A\tau_\nu}). \tag{5.19}$$

If $\tau_\nu \ll 1$, $I_\nu \approx A\tau_\nu B_\nu(T)$ and a linear fit can be used, but this is rarely the case. Skydips are more often used when the atmosphere is just transparent enough to allow observation, so τ is of order unity. $B_\nu(T)(1 - e^{-A\tau_\nu})$ must be fitted to measurements of I_ν as a function of A to estimate τ_ν. Non-linear fitting is required, using numerical techniques.[8]

References

Buscher, D. F., Armstrong, J. T., Hummel, C. A., et al. 1995, ApOpt, 34, 1081

Fried, D. L. 1965, JOSA, 55, 1427

MacEvoy, B. 2013, Astronomical Seeing, Part 1: The Nature of Turbulence, https://www.handprint. com/ASTRO/seeing1.html

Tatarskii, V. I. 1961, Wave Propagation in Turbulent Medium (Berlin: Springer)

---

[8] In practice, a more complicated function is fitted, taking into account the emission of the telescope itself and the frequency response of the system.

Chapter 6

Emission Mechanisms of Electromagnetic Radiation

Understanding the emission mechanisms of photons is crucial to interpreting astronomical observations. Classically, electromagnetic radiation is emitted whenever a charged particle is accelerated, leading to continuum emission. In quantum mechanical systems, particularly atoms and molecules, photons are emitted when a charged particle (usually an electron) transitions between energy levels, leading to spectral-line emission. This chapter will explain the physics of the emission of light.

6.1 Continuum Emission

6.1.1 Bremsstrahlung

How can we accelerate an electron? We could slam it into a positively-charged nucleus. This is known as *bremsstrahlung*, German for *braking radiation*. In situations where the magnetic field is negligible, bremsstrahlung is the most important cooling mechanism of a hot plasma. When a charged particle is slowed (braked) by interaction with an electric field, **E**, the acceleration causes the emission of photons, at the expense of the particle's kinetic energy. The loss of kinetic energy cools the plasma. The electron is free (not bound to a nucleus) before and after the interaction, and hence this is also called *free–free emission*. The process of Bremsstrahlung is shown in Figure 6.1.

The total power of the emitted radiation given by:

$$P_{Br} = \frac{2q^2a^2}{3c^3} \tag{6.1}$$

where q is the electric charge of the particle, and a is its acceleration. The radiation is also polarized with the electric field component vector pointed parallel to the direction of the particle's acceleration. Among the best examples of bremsstrahlung

doi:10.1088/2514-3433/ac087ech6

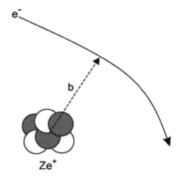

Figure 6.1. Bremsstrahlung is the emission of radiation as a high-speed electron, e^-, passes near a massive nucleus with charge, Ze^+, and is accelerated by the electric field. The impact parameter, b is the closest approach of the electron to the nucleus. Image credit: Wikipedia.

radiation in the universe is that of hot intracluster gas in galaxy clusters. In this case, the electrons are deflected by the Coulomb field of the protons.

Bremsstrahlung radiation is created mainly by a high-speed[1] electron passing by an ionized nucleus of Z protons, having a total charge of Ze^+ coulombs. The impact parameter, b, of the interaction is the closest approach of the electron to the nucleus. Because the acceleration of the electron during the interaction is not uniform, a spectrum of photons is emitted.

The spectrum of bremsstrahlung emission can be determined from the electron's acceleration as a function of time, given by Larmor's[2] formula:

$$P_{Br} = \frac{e^2}{6\pi\varepsilon_0 c^3}v^2. \tag{6.2}$$

The spectrum is flat, up to an upper cutoff frequency, ω_c, which is related to the interaction time, $t = v/b$ or the interaction frequency, $\omega = 1/\delta t = b/v$ (Figure 6.1). The intensity (in Wm^{-2}) of the radiation in the flat part of the spectrum ($\omega < \omega_c$) is given by:

$$I_{Br} = \frac{8Z^2 e^6}{3\pi c^3 m_e^2 v^2 b^2} \tag{6.3}$$

where e is the charge of the electron or proton is in coulombs and m_e is the mass of the electron in kilograms.

Thermal Bremsstrahlung

The discussion above is for a single electron interacting with a single nucleus. In astrophysical situations, bremsstrahlung is created in plasmas containing many electrons and nuclei. We can generalize the above discussion by looking at a large collection of electrons and ions with a thermal distribution at a uniform

---

[1] But for the current discussion, not relativistic, $\frac{1}{2}m_e v^2 \ll m_e c^2$.
[2] Sir Joseph Larmor (1857–1942).

temperature, T. The total emission provided by all the particles in the collection is called *thermal bremsstrahlung* (TB). The distribution of speeds of particles in an ionized cloud with temperature T is given by the Maxwell–Boltzmann velocity distribution: the fraction of particles with speed between v and $v + dv$, per unit dv, is given by:

$$f(v) = 4\pi \frac{m_e}{2\pi kT}^{3/2} v^2 \exp\left[-\frac{m_e v^2}{2kT}\right]. \tag{6.4}$$

In a plasma with free electron number density n_e and ion number density n_i, charged particles will interact with a range of speeds v and impact parameters b. Integrating over the distributions of v and b, we obtain the total power P_{TB} emitted by the plasma cloud, per unit volume:

$$P_{\mathrm{TB}} = g_{ff}(\nu, T)\frac{16\pi e^6}{m_e c^3}\left[\frac{\pi}{m_e kT}\right]^{1/2} Z^2 n_e n_i \exp\left[-\frac{h\nu}{kT}\right] \tag{6.5}$$

where $g_{ff}(\nu, T)$ is the free–free *Gaunt factor*,[3] which is of order unity for a large range in frequency, temperature and density.[4]

6.1.2 Synchrotron Radiation

Magnetic fields accelerate moving particles, and electromagnetic radiation is emitted by an accelerated electrical charge. This form of emission is known as *synchrotron radiation*, named for the radiation emitted from a type of particle accelerator which accelerates electrons by synchronizing pulsed electric and magnetic fields. Charged particles traveling in magnetic field experience a force that is perpendicular to the direction of motion; the particle thus follows a helical path around the magnetic lines of force. The particle will radiate according to the Larmor formula (Equation (6.2)), the relativistic version of which is (Botteon 2019):

$$P = \frac{2q^2\gamma^2}{3m^2 c^3}\left(\frac{dp}{dt}\right)^2$$

where q and m are the charge and mass of the particle respectively, c is the speed of light, $\gamma = \sqrt{\dfrac{1}{1-\frac{v^2}{c^2}}}$, and p is the particle momentum. From the above equation we note that it is the lightest particles such as electrons and positrons, rather than protons, that can be accelerated easily and thus radiate more power. In a non-relativistic regime, the radiation from charges accelerated by magnetic fields is called *cyclotron emission*, but relativistic effects must be considered where the velocity approaches the speed of light; synchrotron radiation can be produced in the form of gamma-rays.

---

[3] In close interactions, quantum mechanical effects play a role. The Gaunt factor is a summation of those effects, thus acting as a correcting factor in this classical formulation.
[4] But one must be careful. Assuming unity may lead to large errors in analysis in a few cases of high-density plasmas.

As electrons enter a magnetic field, the component of their velocity perpendicular to the field lines causes them to spiral along the (generally curvilinear) field lines. The component parallel to the field lines is not affected. If the majority of the radiation comes from the acceleration of spiraling electrons, the radiation is called *synchrotron radiation*. If the majority of the radiation is from the acceleration associated with the motion following the field line, it's called *curvature radiation*. The shape of the spectrum produced (in either case) is determined by the energy distribution of the electrons. The spectrum of the radiation is non-thermal, that is, it looks nothing like blackbody radiation. The radiation is linearly polarized in the plane of the spiral motion, and strongly beamed in the instantaneous direction of the electron's motion (Figure 6.2).

The spectrum is composed of photons emitted by many electrons. If the energy distribution of the electrons obeys a power law, the monochromatic energy flux (units: J s^{-1} m^{-2}; Figure 6.3) is given by:

$$F_\nu = C\nu^\alpha \tag{6.6}$$

where C is independent of frequency, and α is the *spectral index*. There is a *break-over frequency*, ν_b, where the index changes from approximately +2.5 at low frequencies to approximately −1 at high frequencies. The high-frequency index depends on the spectral index of the energy distribution of the electrons. For optically thin radio sources at frequencies relatively near to ν_b:

$$\alpha = \left[\frac{\delta - 1}{2}\right] \tag{6.7}$$

where δ is the spectral index of the energy distribution of the electrons. At low frequencies, the plasma creating the radiation is opaque to its own radiation; this effect is called *synchrotron self-absorption*. The break-over frequency is controlled by the density and temperature of the plasma.

A more practical form of Equation (6.6) is given by observations of the emissions at several frequencies, establishing the value of C_0. Then rewrite Equation (6.6) as:

$$F_\nu = C_0 \left[\frac{\nu}{\nu_b}\right]^{-1} \qquad (\nu > \nu_b) \tag{6.8}$$

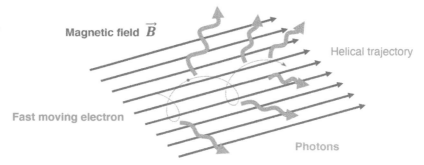

Figure 6.2. Synchrotron emission. The motion of a high-speed electron moving through a magnetic field.

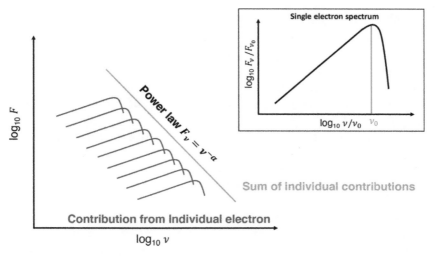

Figure 6.3. The synchrotron emission spectrum follows a power-law decay, and is constructed by adding the contributions from individual electrons.

$$F_\nu = C_0 \left[\frac{\nu}{\nu_b} \right]^{5/2} \qquad (\nu < \nu_b) \qquad (6.9)$$

where C_0 is the measured energy flux at frequency, ν_b. The energy emitted by the plasma between two frequencies ν_1 and ν_2 is then:

$$\varepsilon = \int_{\nu_1}^{\nu_2} F_\nu \, d\nu. \qquad (6.10)$$

Thus, observations of synchrotron radiation aim to establish the values of C, α and ν_b. We may also be able to infer the magnetic field strength in regions of a nebula or plasma, particularly that surrounding a pulsar or neutron star.

An observed range for α is between -3 and $+2.5$ (the theoretical upper limit). Typical spectral index values for various radio sources are:

- Radio galaxy: $\alpha \sim -0.7$
- Pulsar: $\alpha \sim -3$ to -2
- AGN: $\alpha \sim -2$ to $+1$
- SNRs: $\alpha \sim -1$ to -0.35
- PWN: $\alpha \sim -0.35$ to $+0.1$
- PNe: $\alpha \sim -0.3$ to $+2.5$
- H II regions: $\alpha \sim -0.4$ to $+0.5$.

While particularly important to radio astronomers, synchrotron emission can also occur at visible, ultraviolet, X-ray, and gamma-ray frequencies, depending on the energy of the electron and the strength of the magnetic field.

6.1.3 Čerenkov Radiation

Čerenkov[5] radiation (also spelled Cerenkov or Cherenkov) is emitted by particles traveling through a dielectric medium, at a speed that is faster than the velocity of light in that medium. For example, the refractive index of water is $n = 1.33$, meaning the velocity of light in water is $0.75\,c$. Any charged particle (such as an electron) passing through (pure) water with velocity[6] greater than $0.75\,c$ emits Čerenkov radiation.

The effect is analogous to a sonic boom. If the electron were traveling slower than the speed of light, the light emitted by the relaxation of the material's atoms interferes with itself destructively. When the electron is traveling faster than the speed of light, the light emitted interferes constructively, producing intensely energetic emission.

The relationship between the high-speed particle and the emitted radiation can be found by looking at Figure 6.4. The velocity of the particle is v_p and the velocity of the photons is c/n. The angle of emission from the particle path, θ is given by:

$$\cos \theta = \frac{(c/n)t}{v_p t} = \frac{c}{n v_p}. \qquad (6.11)$$

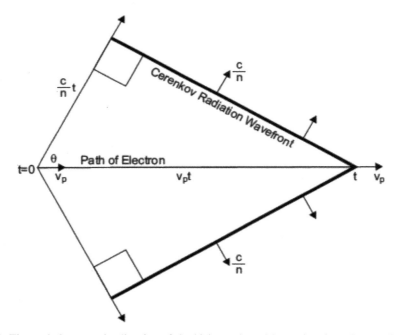

Figure 6.4. The angle between the direction of the high-speed particle passing through a medium and the Čerenkov radiation emitted can be found from the velocities of the particle and the photons.

---

[5] Pavel Čerenkov (1904–1990), Russian physicist, Nobel prize 1958.
[6] This must be the phase velocity, not the group velocity of light in the material.

This angle equals zero when the speed of the incoming particle equals the threshold velocity ($v_p = c/n$), and increases with the speed of the particle. Observations of this angle can be used to measure the velocity (speed and direction) of the incoming particle.

The intensity of the radiation depends on the speed and number of the incoming particles. The spectrum of radiation is continuous, with no peak or break. For frequencies in the optical band, the intensity of the radiation is approximately proportional to the frequency; this explains the bluish glow of nuclear reactors. However, this blue light is seen only with very high-speed particles because Čerenkov radiation is emitted mostly in the ultraviolet band.

There is a frequency above which Equation (6.11) does not hold because n depends on frequency. In the X-ray band, n becomes less than one and so X-rays (or gamma-rays) are not created by the Čerenkov effect.[7]

The Frank–Tamm formula provides the total energy emitted by the medium per unit length of travel of the particle. The process can be viewed as a decay; the change in energy is given by,

$$\frac{dE}{dx} = \frac{q^2}{4\pi} \int_{v_p > c/n(\omega)} \mu(\omega)\omega \left(1 - \frac{c^2}{v_p^2 n(\omega)^2} \right) d\omega \qquad (6.12)$$

where q is the charge of the particle, $\mu(\omega)$ is the permeability and $n(\omega)$ is the index of refraction of the medium (both depend on frequency), v_p is the speed of the particle and ω is the angular frequency of the emitted radiation. The integral is over the spectral region where the speed of the particle is greater than the speed of light in the medium for the particular emission frequency.

Čerenkov radiation is used for observing cosmic rays and neutrinos; examples include the Super-Kamiokande experiment and the Sudbury Neutrino Observatory, as described in Section 7.7.2.

6.1.4 Continuum Emission from Dust

Interstellar dust emits thermal radiation as a modified blackbody spectrum. The temperature of dust grains depends on their size and optical properties, as well as the spectral shape and intensity of the radiation field to which they are exposed in the interstellar medium (ISM). These temperatures usually vary between ~ 2 K and 200 K, but sometimes temperatures as high as 2000 K are reached, resulting in sublimation if the dust grains are in close proximity to a powerful radiation source. Equilibrium temperatures can be exceeded when hard-UV photons are absorbed by small grains (Draine & Li 2001). In the case of a region with a weak interstellar radiation field, the temperature of the dust grains will be lower. Similarly, the temperature will also be lower if dust grains are shielded from a radiation source. But in both these cases, the temperatures will exceed the (CMB) temperature. If the

---

[7] In fact, there are a few special energies where X-rays may be created; we will neglect these in what follows.

dust grain temperatures are not higher than the CMB temperature, the dust emission from distant galaxies cannot be detected.

The Emission Spectrum, Dust Mass and Temperature

Temperature T_d and emissivity ε_ν are the most crucial quantities for dust emission. As discussed, the temperature of the dust grains will vary within a galaxy. If the emitting region is be optically thin, the emission spectrum, f_ν can be expressed by $f_\nu \propto B_\nu \varepsilon_\nu$, where B_ν is the Planck function, and the emissivity ε_ν quantifies the departure from a perfect blackbody emitter.

In the presence of absorption, the emitted spectrum is modified according to the optical depth τ_ν,

$$f_\nu \propto [1 - \exp(-\tau_\nu)]B_\nu. \tag{6.13}$$

The peak of f_ν approximates a blackbody spectrum when the opacity near a wavelength of 100 micrometers is large. This causes the emission in the f_ν peak to be suppressed, corresponding to the Rayleigh–Jeans regime. This f_ν provides a good fit at higher values of temperature for submillimeter and far-IR data. As can be seen in Figure 6.5, there are fewer than four data points for the majority of observed Spectral Energy Distribution (SED)s for high-redshift galaxies. Thus, the effect of τ_ν may be not significant.

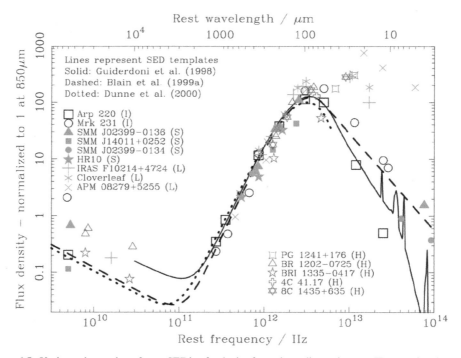

Figure 6.5. Various observed restframe SED's of galaxies from the radio to the near-IR wave bands. Image credit: Reprinted from Blain et al. (2002), Copyright (2002), with permission from Elsevier.

For the mid-IR, we can reasonably assume that the low-frequency part of the spectrum will resemble a modified blackbody function. At higher frequencies, a power law $f_\nu \propto \nu^\alpha$ is a better fit to the SED than a Wien spectrum exponential drop-off (see Figures 2.6 and 6.5). This can be explained by both shorter wavelength emission from hotter dust components and near-IR wavelength stellar emission. The absence of an exponential drop-off is also exhibited in Arp 22 and Mrk 231 in Figure 6.5.

In order to explore the star formation history of a galaxy, it is useful to measure the metal content of the ISM. A census of the metals, in turn, requires a measurement of the dust mass of the galaxy M_d (Hughes et al. 1997; Omont et al. 2001). Conventionally, for this purpose, a mass absorption coefficient κ_ν is used, which is ν dependent and is proportional to ε_ν (Draine & Lee 1984). Units for κ_ν are $m^2\ kg^{-1}$ and is a measure of the cross section area per unit mass for blackbody emission. This mass absorption coefficient relates dust mass of a galaxy to its SED f_ν and luminosity L,

$$\kappa_\nu B_\nu M_d = L \frac{f_\nu}{4\pi \int f_{\nu'}\, d\nu'}. \tag{6.14}$$

Assuming a spherical dust grain with density ρ, radius a and an emission cross section of πa^2, the cross section per unit mass is $\kappa_\nu = \pi a^2/(4/3\pi a^3 \rho) = 3/4a\rho$. For non-spherical grains, we can define a dimensionless factor Q_ν via $\kappa_\nu = 3Q_\nu/4a\rho$ (Hildebrand 1983). Dust grains are mostly of irregular shapes, as shown in Figure 6.6. Then $Q_\nu B_\nu$ is the energy flux per unit area of a grain surface; it is effectively an emissivity function (see https://ned.ipac.caltech.edu/level5/Sept04/Blain/Blain2_2.html). However, knowing the appropriate Q_ν factor for a given type of dust is non-trivial.

For submillimeter data, in addition to the limitations above, the mass of a galaxy cannot be estimated easily without determining its dust temperature. A galaxy's flux

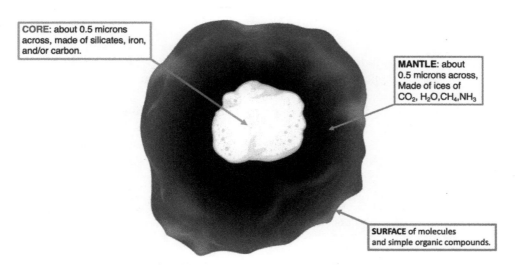

CORE: about 0.5 microns across, made of silicates, iron, and/or carbon.

MANTLE: about 0.5 microns across, Made of ices of CO_2, H_2O, CH_4, NH_3

SURFACE of molecules and simple organic compounds.

Figure 6.6. The figure shows a model of a typical dust grain.

density S_ν is directly proportional to its T_d in the Rayleigh–Jeans regime, $S_\nu \propto \nu^{2+\beta} M_d T_d$. Dust mass can be estimated from a single long-wavelength observation. As $L \propto S_\nu T_d^{3+\beta}$, any uncertainty factor in T_d will produce the same factor of uncertainty in dust mass, and a higher level of uncertainty in luminosity.

It is not known whether the procedure used to determine dust mass of nearby galaxies, where they happen to produce reliable results, may be used to determine dust mass in more distant and luminous galaxies. Due to the lack of high-quality data for high-redshift galaxies in submillimeter and far-IR wave bands, over-parameterization and over-interpretation of observations of their SEDs is a constant danger.

6.2 Spectral-line Emission

When a quantum system (such as an atom or molecule) changes its state, it moves from one specific energy state to another. This *transition* will itself have a specific energy. If that energy is gained or lost by electromagnetic radiation, then the electromagnetic radiation will have a specific frequency, defined by Planck's law: $\nu = E/h$.

When electromagnetic radiation is analyzed by frequency (or color or wavelength or energy), there will therefore be certain frequencies that stand out from the broader continuum radiation, either as brighter or fainter. Because the light at these frequencies is directly related to the quantum processes happening in and around the material, they can be used to probe not only the makeup of that material, but also the physical conditions in and around it. Spectroscopy in its many forms is therefore a fundamental tool of astrophysics.

These brighter or darker frequencies are *spectral lines*, so called because in optical spectra they appear as bright or dark lines perpendicular to the direction of dispersion of the light. They were first observed in an astronomical context as dark lines in the solar spectrum by Wollaston[8] and Fraunhofer,[9] and are still known as the *Fraunhofer lines*.

Spectral lines were seen in emission and absorption in the laboratory by physicists and chemists including Ångström, Alter,[10] Kirchhoff, Bunsen[11] and others. Based on these laboratory studies, Kirchhoff published his three laws of radiation in 1859:

1. A hot solid, liquid, or gas, under high pressure, gives off a continuous spectrum, now known as blackbody radiation.
2. A hot gas under low pressure produces a bright-line or emission line spectrum.
3. A dark-line or absorption spectrum is seen when a continuous source to viewed through a cool gas under pressure.

---

[8] William Hyde Wollaston (1766–1828), English physicist and chemist, who worked extensively with platinum and discovered palladium and rhodium. He thought the dark lines were the "edges" of the primary colors.
[9] Joseph von Fraunhofer (1787–1826), German optician.
[10] David Alter (1807–1881), American inventor and scientist, was the first to propose that every element may have its own emission spectrum, in 1854.
[11] Robert Bunsen (1811–1899), German chemist, studied the emission spectra of burning elements, discovered (with Kirchhoff, 1861) caesium and rubidium in the solar spectrum, and invented the Bunsen burner.

The second and third laws are related: the wavelengths of the emitted lines from the hot gas and the absorbed lines in the cool gas are identical. With greater understanding of the microphysics, we can now say that a spectral line may be seen in *emission*, where, for that particular transition, the matter (e.g., atom) is moving from the higher-energy to the lower-energy state and emitting a photon; or in *absorption*, where, for the same transition, the matter is moving from lower-energy to higher-energy, absorbing a photon of the same frequency to provide the required additional energy. While this understanding was first achieved for the spectra of atomic gases and plasmas, it is generally true for all spectral lines.

Permitted and Forbidden Lines

As well as a characteristic frequency, a spectral line has greater or lesser *line strength*. A stronger line will be brighter in emission, and darker in absorption. The brightness or darkness of the line is largely defined by the rate at which photons are emitted or absorbed. The rate of photon emission or absorption is given by the number of atoms or molecules that are in the correct state to emit or absorb (which is defined by the physical conditions of the gas—this makes spectral lines good astrophysical probes), multiplied by the probability of the transition. This transition probability is characteristic of the quantum transition, and is in turn related to the *Einstein coefficients*.

The largest Einstein coefficients, and hence strongest spectral lines, come from *electric dipole* transitions. Atoms or molecules that are in excited states, and which can decay by an electric dipole transition, will generally do so in nanoseconds. These *permitted transitions* give rise to the dominant features of laboratory spectra, termed the *permitted lines*. Permitted transitions are those that obey a certain set of quantum *selection rules*.

Different sets of selection rules generate different transitions. *Magnetic dipole* transitions are weaker than electric dipole ones, and electric quadrupole and magnetic quadrupole transitions weaker still. These transitions are termed *forbidden transitions*: in laboratory spectra, atoms or molecules in excited states that can only decay through forbidden transitions will generally decay due to their collision with another particle (*collisional de-excitation*) before they decay by forbidden transition, so the *forbidden lines* are effectively invisible. In the much lower densities of interstellar space, collisions are much rarer, and excited atoms or molecules may be able to decay by emitting a photon before collisional de-excitation. In this case, the spectrum of e.g., a nebula may be dominated by forbidden lines.

6.2.1 Spectral-line Emission Mechanisms

Electronic Transitions

The spectral lines identified in the 19th century—and still the workhorses of astrophysical spectroscopy—arise from the electrons in atoms gaining and losing energy as they move between higher- and lower-energy orbitals. This process is governed by quantum mechanics, and generates spectral lines stretching from the radio to the X-ray portions of the electromagnetic spectrum.

Hydrogen and Similar Atoms: Hydrogen—one proton and one electron—is the simplest of the atoms. It is a two-body problem, and the quantum mechanical equations governing the interaction of the two particles can be solved. The spectrum of hydrogen (i.e., the set of all spectral lines arising from it) was understood phenomenologically before the development of quantum theory provided a physical explanation. Balmer[12] found that the wavelengths of optical hydrogen lines followed a formula:

$$\frac{1}{\lambda} = R_H\left[\frac{1}{2^2} - \frac{1}{n^2}\right] \tag{6.15}$$

where n (now known to be the principal quantum number, representing the electron's energy level) is an integer greater than 2 and $R_H = 10{,}967{,}758.306 \pm 0.013$ m^{-1} is the hydrogen Rydberg.[13] The lines whose wavelengths are predicted by this formula are known as the Balmer lines (see e.g., Table 6.1). Since n can take values up to infinity, an unlimited number of Balmer lines exist; over 30 have been observed, and 6 of them are the only hydrogen lines in the optical region of the spectrum, making them so important to astronomical spectroscopy that they are known as the "hydrogen lines." The $n = 2$ to $n = 3$ transition ($2 \rightarrow 3$) is known as hydrogen-alpha (Hα), the $2 \rightarrow 4$ transition is Hβ, and so on. (Though the Greek letter designation is used only for the first few transitions, after which they are numbered.)

The full spectrum of hydrogen consists of a long procession of line series like the Balmer lines. Each series starts with a first line (e.g., Hα) at the lowest frequency, with subsequent lines at higher frequencies and decreasing frequency gaps between them, until the highest-frequency lines blend together at a high-frequency limit. All these spectral lines can be characterized by the Rydberg formula, which is a generalization of Balmer's formula:

$$\frac{1}{\lambda} = R_H\left[\frac{1}{n_1^2} - \frac{1}{n_2^2}\right] \tag{6.16}$$

where $n_1 = 1, 2, 3, \ldots$ and $n_2 > n_1$. Each series of spectral lines corresponds to a particular value of n_1, and the first six are named: Lyman,[14] Balmer, Paschen,[15] Brackett,[16] Pfund,[17] and Humphreys[18] (Table 6.1).

Bohr used his orbital model of the atom to give a physical explanation of the Rydberg formula, in which n_1 corresponds to the lower-energy electron orbital state, and n_2 to the higher-energy state. The first, lowest frequency, line in the series corresponds to the electron's transition from n_1 to $n_2 = n_1 + 1$, and the high-

---

[12] Johann Jakob Balmer (1825–1898), Swiss mathematician.
[13] Johannes Robert Rydberg (1854–1919), Swedish physicist.
[14] Theodore Lyman (1874–1954), American physicist.
[15] Friedrich Paschen (1865–1947), German physicist.
[16] Frederick Brackett (1896–1988), American physicist.
[17] August Pfund (1879–1949), German–American physicist.
[18] Curtis Judson Humphreys (1898–1986), American physicist.

Table 6.1. Frequencies and Wavelengths of the Low-frequency Ends of the First Six Spectral Series for Hydrogen

Series	$n_1 \rightarrow n_2$	ν(THz)	λ(nm)	Region	Symbol
Lyman	$1 \rightarrow 2$	2467	121.6	Ultraviolet	Lyα
	$1 \rightarrow 3$	2923	102.6	Ultraviolet	Lyβ
	$1 \rightarrow 4$	3085	97.5	Ultraviolet	Lyγ
	$1 \rightarrow 5$	3159	94.97	Ultraviolet	Lyδ
	$1 \rightarrow 6$	3199	93.78	Ultraviolet	Lyε
Balmer	$2 \rightarrow 3$	456	656.5	Optical	Hα
	$2 \rightarrow 4$	617	486.3	Optical	Hβ
	$2 \rightarrow 5$	690	434.2	Optical	Hγ
	$2 \rightarrow 6$	731	410.3	Optical	Hδ
	$2 \rightarrow 7$	755	397.1	Optical	Hε
Paschen	$3 \rightarrow 4$	160	1876	Infrared	Pα
	$3 \rightarrow 5$	234	1282	Infrared	Pβ
	$3 \rightarrow 6$	274	1094	Infrared	Pγ
	$3 \rightarrow 7$	296	1005	Infrared	Pδ
	$3 \rightarrow 8$	314	954.9	Infrared	Pε
Brackett	$4 \rightarrow 5$	74.0	4052	Infrared	Brα
	$4 \rightarrow 6$	114	2626	Infrared	Brβ
	$4 \rightarrow 7$	139	2166	Infrared	Brγ
	$4 \rightarrow 8$	154	1945	Infrared	Brδ
	$4 \rightarrow 9$	165	1818	Infrared	Brε
Pfund	$5 \rightarrow 6$	40.2	7460	Infrared	Pfα
	$5 \rightarrow 7$	64.5	4653	Infrared	Pfβ
	$5 \rightarrow 8$	80.2	3740	Infrared	Pfγ
	$5 \rightarrow 9$	91.0	3297	Infrared	Pfδ
	$5 \rightarrow 10$	98.7	3039	Infrared	Pfε
Humphreys	$6 \rightarrow 7$	24.2	12,370	Infrared	Huα
	$6 \rightarrow 8$	40.0	7503	Infrared	Huβ
	$6 \rightarrow 9$	50.8	5908	Infrared	Huγ
	$6 \rightarrow 10$	58.5	5129	Infrared	Huδ
	$6 \rightarrow 11$	64.2	4673	Infrared	Huε

Note. The most commonly observed transitions in astronomy are Hα, Hβ, and Brγ.

frequency limit to the electron's move out to infinity, i.e., ionization. The energy required to move the electron from its lowest state ($n_1 = 1$) to infinity is therefore the *binding energy*, W, of the electron, and is related to the Rydberg constant:

$$R = \frac{W}{hc}. \tag{6.17}$$

The binding energy of hydrogen is 13.6 eV, corresponding to a wavelength of 912 Å, which is the high-frequency limit of the Lyman series (Table 6.1). Thus, electro-magnetic radiation with $\lambda < 912$ Å (the extreme ultraviolet or EUV) will be absorbed by hydrogen atoms, which in turn will be ionized; EUV observations are thus limited to a local bubble of low hydrogen density in the interstellar medium.

The Rydberg formula can be applied to atoms and ions other than hydrogen, as long as they follow the model of an electron orbiting a positive nucleus. A *hydrogenic ion* is one where all but one of the original atom's electrons have been removed (e.g., He^+). The formula is altered to take account of the higher positive charge of the nucleus. A *Rydberg atom* is one in which a lone outer electron orbits a nucleus whose positive charge is largely screened by inner electrons, leaving a net charge approximating one proton, allowing the Rydberg formula to be used.

The hydrogen spectrum continues through the infrared and into the radio, where the spectral lines are known as *radio recombination lines*, and are generally associated with the recombination of ionized hydrogen. They are generally written as e.g., H92α, where $n_1 = 92$ and $n_2 = 93$.

Complex Atoms: Once the atomic system consists of a nucleus and more than one electron, the quantum mechanical equations describing the energy states of the electrons become a many-body problem and are not solvable in closed form—and the more electrons, the more possible energy level combinations.

Electron configurations in atoms are defined by the Pauli exclusion principle,[19] that no two electrons can have the same set of quantum numbers. As the number of electrons grow, extra electrons are forced to occupy higher-energy quantum states. The quantum numbers that define the states available to electrons are:

1. The principal quantum number, $n = 1, 2, 3...$, as used in the Rydberg formula. Each value of n defines an electronic shell around the nucleus.
2. The azimuthal quantum number, $l = 0, \pm 1, ..., \pm(n - 1)$, defines an increasing number of subshells with larger n, denoted by letters s, p, d and f.
3. The magnetic quantum number, $m_l = 0, \pm 1, ..., \pm(l)$ defines orbitals within the subshell, with an increasing number of orbitals available for higher l.
4. The electron spin quantum number, $m_s = -1/2, +1/2$ for each orbital.

The Pauli exclusion principle thus implies that each orbital can hold up to two electrons, and further electrons must go into higher orbitals. When orbitals have the same energy (called *degenerate*), the electrons minimize their energy by filling the orbitals with one electron per orbital with aligned spins before adding second electrons into the orbitals (Hund's[20] rules). Table 6.2 shows the quantum numbers and numbers of electrons associated with the various subshells, and it is this proliferation of available electronic states that gives the periodic table its shape (Figure 6.7).

The electronic states of atoms and ions are written in spectroscopic notation, which summarizes the states of all the electrons in the atomic system. This notation

---

[19] Wolfgang Pauli (1900–1958), German–American–Swiss physicist.
[20] Friedrich Hund (1896–1997), German physicist.

Table 6.2. Possible Values of the Quantum Numbers

Shell n	Subshell l	Subshell Designation	Magnetic m_l	Total Orbitals	Total Electrons
1	0	1s	0	1	2
2	0	2s	0	1	2
	1	2p	−1, 0, +1	4	8
3	0	3s	0	1	2
	1	3p	−1, 0, +1	4	8
	2	3d	−2, −1, 0, +1, +2	9	18
4	0	4s	0	1	2
	1	4p	−1, 0, +1	4	8
	2	4d	−2, −1, 0, +1, +2	9	18
	3	4f	−3, −2, −1, 0, +1, +2,+3	16	32

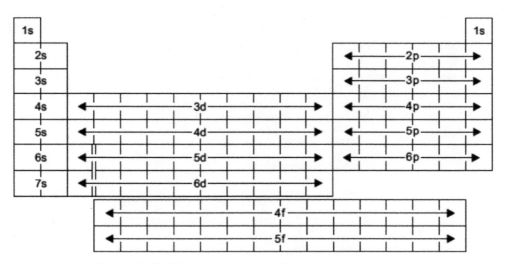

Figure 6.7. The filling of electron subshells defines the periodic table.

defines the system by the set of values $[n, l, S, L, J]$; n and l are the principal and azimuthal quantum numbers as above, and are used to define the electronic state of the highest populated orbital. For example, a neutral carbon atom has 6 electrons—4 of them fill the 1s and 2s subshells, and the remaining 2 are in the 2p subshell, so the first part of the spectroscopic designation is $2p^2$, ignoring the filled subshells. (This first part is often left out.) The second part—the term symbol—describes the total angular momentum of the electronic system, using quantum numbers that refer to the vector sums of the spin (S) and orbital (L) angular momenta of the electrons, and the vector sum of the vector sums of spin and orbital angular momenta for each electron (J). For light atoms (generally the ones of interest to astrophysicists), $J = L + S$ (Russell–Saunders coupling). The term symbol is written $^{(2S+1)}L_J^{\pi}$, where L is given as a letter symbol rather than a number, and π is the parity, expressed as 0 for odd parity, and left out otherwise. In the case of the neutral carbon atom, $S = \Sigma m_s = 1$ (the spins of the paired electrons in the s subshells cancel out, leaving the two unpaired electrons in separate p-subshell orbitals, with parallel spins, summing to 1). $L = \Sigma m_l = 1$, as the s orbitals have $m_l = 0$ and the 2 outer electrons will be found in orbitals with $m_l = 0$ and $m_l = 1$. $L = 1$ is written as P, by analogy with the subshell designation. J can take values of 0, 1, 2. The term symbol is then $^3P_{(0,1,2)}$. According to Hund's rules, in this case, the $^3P_0$ state is the lowest-energy and hence the ground state. Transitions between the $J = 0$, $J = 1$, and $J = 2$ states give rise to *fine-structure lines* observed in the submillimeter-wave spectrum. The superscript 3 denotes a *triplet* state; 1 and 2 denote *singlet* and *doublet*, respectively (Tables 6.3 and 6.4).

Ionized Atoms: When an atom is ionized, losing one or more electrons, its electronic structure, and hence its spectrum, changes. Atoms with similar electronic structures (such as members of the same group in the periodic table) have similarities in their spectra, so an ionized atom will have a spectrum more like a different atom altogether.

The ionization state in spectroscopy is denoted by Roman numerals: the neutral state is I (e.g., HI or CI for neutral hydrogen or carbon). As the atom loses electrons, the Roman numeral is increased: HII is singly ionized hydrogen. Technically, this notation applies to the spectrum of the atom, rather than the atom itself, so that the

Table 6.3. Nomenclature and Symbols Used in Spectroscopy

Symbol	Meaning
n	Principal quantum number
	$n = 1, 2, 3, 4, 5, 6...$
	shell K L M N O P $\cdots$
l	Orbital angular momentum of individual electron
	$l = 0, 1, 2, 3, 4, 5, ...$
	subshell s p d f g h $\cdots$
L	Total orbital angular momentum
$L = \sum(l)$	$L = 0, 1, 2, 3, 4, 5 ...$
	S P D F G H $\cdots$
s	Angular momentum of an individual electron
$S = \sum(s)$	Total spin angular momentum
$j = l + s$	Spin–orbit coupling for single electron
$J = \sum j$	(jj coupling) Total angular momentum, atoms above iron
$J = L + S$	(LS coupling) Total angular momentum, atoms below iron
M	Magnetic quantum number, or the z-components of J along an external magnetic field

Table 6.4. Interpreting the Spectroscopic Notation of an Atomic Energy Level

Part:	Example energy state: $2p^3\ {}^4S^{\circ}_{3/2}$ Meaning
2 :	The shell, $n = 2 \to L$ shell.
p^3 :	The subshell, $l = 1 \to p$ subshell, with three outer (valence) electrons.
	The term:
4 :	The multiplicity, or the degeneracy of the orbital. With spin $S = 3/2$, (the sum of the spins of the three outer electrons) multiplicity $= 2S + 1 = 4$.
S :	The total angular momentum of the state, $L = 0 \to S$ orbital.
3/2 :	The total angular momentum, $J = L + S = 0 + 3/2 = 3/2$. This part is included to account for relativistic effects.
∘ :	Odd parity. There is usually no superscript for even-parity terms
	The inclusion of M (the Zeeman effect) will split terms into states, with $M = -J, \ldots, +J$.

OIII spectrum arises from the O^{2+} ion. Astronomers, however, routinely use the spectroscopic designation to refer to the atoms and ions that generate the spectra—so a volume of ionized hydrogen, H^+, is referred to as an HII region.

Although the spectrum of an ion will be similar to that of an atom with the same electronic structure, the ion will have a nucleus with its original positive charge, greater than that of the neutral analog. This shifts the spectral features of the ion to shorter wavelength. The electronic transitions of highly-ionized heavy atoms can have energies in keV (compared to the 13.6 eV Lyman limit of hydrogen) and these transitions give rise to X-ray spectral features.

Selection Rules: Permitted and forbidden transitions of various types are possible as electrons change their quantum state. The selection rules that define the electric dipole (E1), magnetic dipole (M2) or quadrupole (M3), or electric quadrupole (E2) transitions are listed in Table 6.5. The selection rules operate on the angular momentum quantum numbers, and also on the *parity* π of the electron state, determined by reflecting the electron wave function through the nucleus. If the orbital angular momentum quantum numbers (m) of the electrons are summed:

$$\pi = (-1)^{\sum_i m_i}. \tag{6.18}$$

Nucleon Transitions

The nucleons (protons and neutrons) that make up an atomic nucleus populate nuclear orbitals that are analogous to the electronic orbitals. Excited nuclei have nucleons in higher-energy orbitals than their ground state; as the nucleon drops to a ground-state orbital, a photon is emitted. Since nuclear binding energies are so much greater than electronic binding energies, the nuclear de-excitation transitions have much higher energy, generally in the MeV range. These transitions give rise to gamma-ray spectral lines.

Table 6.5. Selection Rules for Atomic Electron Transitions

Rule	E1, E Dipole "Permitted"	M2, M Dipole "Forbidden"	E2, E Quadrupole "Forbidden"	M3, M Quadrupole "Forbidden"
1	$\Delta J = 0, \pm 1 (0 \nrightarrow 0)$	$\Delta J = 0, \pm 1 (0 \nrightarrow 0)$	$\Delta J = 0, \pm 1, \pm 2$ $(0 \nrightarrow 0, 1, \frac{1}{2} \nrightarrow \frac{1}{2})$	$\Delta J = 0, \pm 1, \pm 2$ $(0 \nrightarrow 0, 1, \frac{1}{2} \nrightarrow \frac{1}{2})$
2	$\Delta M_J = 0, \pm 1$ If $\Delta J = 0$ $0 \nrightarrow 0$	$\Delta M_J = 0, \pm 1$ If $\Delta J = 0$ $0 \nrightarrow 0$	$\Delta M_J = 0, \pm 1, \pm 2$	$\Delta M_J = 0, \pm 1 \pm 2$
3	$\pi_f = -\pi_i$	$\pi_f = \pi_i$	$\pi_f = \pi_i$	$\pi_f = -\pi_i$
4	If $\Delta S = \pm 1$ $\Delta L = 0, \pm 1, \pm 2$	If $\Delta S = \pm 1$ $\Delta L = 0, \pm 1, \pm 2$	If $\Delta S = \pm 1$ $\Delta L = 0, \pm 1, \pm 2, \pm 3 (0 \nrightarrow 0)$	If $\Delta S = \pm 1$ $\Delta L = 0, \pm 1 (0 \nrightarrow 0)$
5	One e^- jump. $\Delta l = \pm 1$	No e^- jump. $\Delta l = 0, \Delta n = 0$	No or one e^- jump. $\Delta l = 0, \pm 2$	One e^- jump. $\Delta l = \pm 1$
6	If $\Delta S = 0$ $\Delta L = 0, \pm 1$ $0 \nrightarrow 0$	If $\Delta S = 0$ $\Delta L = 0$	If $\Delta S = 0$ $\Delta L = 0, \pm 1, \pm 2$ $0 \nrightarrow 0, 1$	If $\Delta S = 0$ $\Delta L = 0, \pm 1, \pm 2$ $0 \nrightarrow 0, 1$

Notes. Rules 1, 2, and 3 are rigorous—they must always be followed. Rule 4 applies when there is negligible configuration interaction, rules 5 and 6 apply when LS coupling is valid. From left to right, transitions following each column or rules have probabilities several orders of magnitude lower than the previous one.

Molecular Transitions

Molecules are quantum systems like atoms and nuclei, albeit more complicated. They consist of two or more atoms that share pairs of electrons, resulting in a lower overall energy than the atoms would have on their own. This binding energy is much lower than that of electrons in atoms or of nucleons in a nucleus, so molecules are generally found in lower-temperature environments, where there is not enough energy available to break up the molecule. Molecules can also be broken up by *photodissociation* by UV photons, and so they are also often found in darker environments, where UV light has been blocked. Molecules are found in dense dark interstellar gas clouds, in shells around dying giant stars, and in the dusty disks around the centers of galaxies. Simple molecules (such as TiO) are also found in the atmospheres of cool red stars (particularly M stars, where they dominate the optical spectrum).

The complexities of molecular quantum mechanics mean that it is generally not possible to calculate their transitions precisely—approximations are used, and generally verified by measurement. The Born–Oppenheimer approximation relies on the large mass of the nuclei compared to the mass of the electrons. The nuclei are considered to be slow-moving, and so a static nuclear potential can be calculated. This defines the possible electron states, thus the possible quantum states of the molecule, and thus the possible transitions between such states.

Rotational Transitions: Because a molecule has a spatially-extended shape, it can rotate, and its rotation is quantized, giving rise to transitions between states of greater or less angular momentum. These rotational transitions generally have rather low energy, and so are mainly seen in the radio-to-terahertz spectrum.

The simplest rotational spectra are those of diatomic molecules, defined by quantum number $J = 0, 1, 2...$, with energy levels

$$E_{rot} \approx \frac{\hbar^2}{2\mu D^2} J(J + 1) \tag{6.19}$$

where $\mu = m_1 m_2/(m_1 + m_2)$ is the reduced mass of the molecule, D is the equilibrium distance between the nuclei, and the moment of inertia of the molecule is μD^2. The rotational constant is then defined as $B = \hbar/(4\pi\mu D^2)$, and the frequencies of the rotational transitions with $\Delta J = 1$ are given by

$$\nu(J) \approx 2B(J + 1). \tag{6.20}$$

This approximation does not take into account the fact that, as the molecule rotates faster, it stretches out, changing the moment of inertia.

The most commonly observed diatomic molecule is carbon monoxide, CO, and its rotational constant is 57.6 GHz, giving rotational spectral lines at frequencies close to 115, 230, 345, 460 $\cdots$ GHz.

To have an electric dipole (i.e., permitted) $\Delta J = 1$ transition, the molecule must have an electric dipole moment. *Homonuclear* diatomic molecules (i.e., two identical atoms such as H_2 or O_2) have no dipole, so their rotational transitions are not permitted. *Heteronuclear* diatomic molecules, on the other hand, in which the atoms are dissimilar (e.g., CO, CN, OH), have a permanent dipole and permitted transitions.

Polyatomic molecules (those with more than two atoms) can have similar spectra to diatomic ones, if their atoms are arranged close to a straight line. Examples of this are the cyanopolyyne family (HCN, HC_3N, HC_5N, HC_7N, ...) and the HCO^+ molecular ion.

Most polyatomic molecules, however, will have a more complex shape, and far more complex spectra, than linear molecules. Many of them, however, have axes of symmetry in their structure and are known as *symmetric top* molecules. For some molecules, such as ammonia (NH_3) and formaldehyde (H_2CO), the effect of this symmetry is to create multiple "ladders" of energy levels, with permitted transitions inside the ladder, and forbidden transitions between ladders. While the relative populations of the levels in a ladder are largely defined by the emission and absorption of photons, the relative populations of states in different ladders are largely defined by collisions between molecules, which in turn depends on the *kinetic temperature* of the gas, a measure of the particle speed. Analysis of multiple spectral lines of these molecules may therefore be used to estimate the physical temperature of the gas giving rise to the spectral lines.

Vibrational Transitions: The vibrational energy of a molecule is also quantized. Vibrational states have greater energy than rotational states, and each vibrational state will have its own rotational spectrum. (Rotational transitions within a vibrationally-excited state can often be seen in the shells of material around old stars, such as IRC+10216.) Transitions between vibrational states are generally found in the infrared spectrum, and, because the excited states require high energy, their environments are usually quite hot. Vibration transitions often include a change of rotational state as well, and may be called ro-vibrational transitions.

One of the best-observed examples is a quadrupole transition of molecular hydrogen, H_2. This transition can be observed at 2.122 μm wavelength in the atmospheric K-band window. The transition is labeled 1–0 S(1), meaning that the vibration state drops from $v = 1$ to $v = 0$, and the rotation state from $J = 3$ to $J = 1$. This emission can be fluorescent, in which case the molecules have been excited by UV radiation, or shock-excited, where the molecules have been heated by the passage of an interstellar shock (Burton 1992).

Other Transitions: More complicated molecules can move in complicated ways, such as bending, stretching, and "wagging." These all produce energy states with transitions, and hence spectral lines, but the spectra are very complicated. Frequently, many spectral lines blend into each other, forming a band, and they are usually found in the infrared.

Some molecules have low-frequency transitions in which the symmetry of the molecule changes. An example of this is the inversion transition of ammonia (NH_3), in which the hydrogen atoms define a plane with the nitrogen atom above or below it. The transition between the possible nitrogen positions gives rise to a radio spectral line at 1.3 cm (~23 GHz).

Isotopically-substituted Molecules: The different isotopes of various elements are chemically identical, having the same number and configuration of electrons. It is therefore possible to have molecules of the same chemical species containing different isotopes. For example, the CO molecule is usually comprized of ^{12}C and ^{16}O, the most common isotopes; but it is possible to have ^{13}CO, in which the carbon atom is carbon-13 rather than carbon-12. This type of molecule is referred to as *istopically-substituted* or as an *isotopologue* (*isotopomer* is an older term). Isotopically-substituted molecules have slightly different spectra, due to the difference in mass of the nuclei. For example, ^{13}CO is heavier that the normal version, and so has a larger moment of inertia, a smaller rotational constant, and spectral lines with slightly lower frequencies (110 GHz compared to 115 GHz).

Laser and Maser Emission

Spectral lines can exhibit laser or maser behavior, if they have *population inversions*. In a normal thermodynamic system, the higher-energy (upper) level of a transition will have a smaller population than the lower-energy (lower) level; in a population inversion, the upper level has a higher population than the lower one. This anomalous population structure leads to the *stimulated emission* of the transition becoming dominant.

Einstein found that, in order to be consistent with Planck's and Kirchhoff's laws, atoms or molecules distributed across and upper and lower level must interact with radiation through absorption, emission, and stimulated emission. The transition probabilities of these three processes are defined by the *Einstein coefficients*, A_{ul}, B_{lu}, and B_{ul}, where u and l refer to the upper and lower states. A_{ul} is the probability per second that any given atom or molecule will decay from higher to lower state, so $N_u A_{ul}$ is the number of photons spontaneously emitted per unit volume per second, where N_u is the volume density of atoms or molecules in that state (and N_l is density in the lower state). Absorption requires radiation to be absorbed, so depends on the amount of incoming radiation: the number of absorbed photons per second per unit volume is $N_l B_{lu} \bar{U}$, where $\bar{U}$ is the average energy density of the electromagnetic radiation. The number of photons emitted per second per unit volume due to stimulated emission is also dependent on the radiation energy density: $N_u B_{ul} \bar{U}$.

The two Einstein B coefficients are similar in size, so stimulated emission will generally be reabsorbed. When there is a population inversion, however, the stimulated emission will be greater than the absorption. This allows a positive-feedback loop in which spontaneous emission triggers stimulated emission, which triggers more stimulated emission, the overall emission being amplified as the the radiation travels through the material; such a system is a maser (or a laser if the radiation is of infrared wavelength or shorter), and can be characterized by a negative optical depth. Under normal circumstances, this effect will swiftly deplete the upper level population. A maser relies on *pumping*, in which more excited atoms or molecules decay into the upper level of the maser transition faster than they can decay out of it by collision or spontaneous emission. A large population then builds up in the upper level, which maintains the population inversion. Maser pumping may be *collisional* (initial excitation by collision between particles) or *radiative* (initial excitation by absorption of higher-frequency radiation, often infrared).

Observed astrophysical masers are generally molecular transitions, and the most commonly observed molecules are hydroxyl (OH), water (H_2O), methanol (CH_3OH), hydrogen cyanide (HCN), and silicon monoxide (SiO). They are useful to astrophysics in two main ways, as beacons and as physical probes. Because masers are very bright in their specific spectral line, they are easy to see, and therefore serve as signposts of a particular set of astrophysical conditions that generate the population inversion. In particular, regions of star formation often generate OH, H_2O, and CH_3OH masers, while the shells around old, evolved stars often generate OH, H_2O, and SiO masers.

The very high brightness of maser lines also makes it easy to measure their frequency and frequency width. This in turn allows the precise measurement of environmental effects that alter these quantities. In particular, the Zeeman effect (see below) can be used to estimate the local magnetic field, and the Doppler shift can be used to give accurate velocity measurements. Velocity measurements of masers in the circumnuclear disks of galaxies have been used to trace Keplerian rotation around the supermassive black hole in these galaxies, and hence derive the mass of the black hole (Miyoshi et al. 1995).

Maser sources tend to be very small in spatial extent—since they rely on quite specific physical conditions, which may only occur in small volumes—so they are most effectively observed with high angular resolution. Although masers can be seen with single-dish radio telescopes (and widefield surveys to find masers often use single dishes) they are generally observed with radio interferometers, often with VLBI (Section 1.3.3). This very small size has been used to determine accurate distances to masers by parallax (the shift in apparent position of a celestial source as the Earth orbits the Sun), and hence to accurately measure the Milky Way (Reid et al. 2016).

6.2.2 Effects on Spectral Lines

The physical conditions of the medium giving rise to spectral lines can have significant effects on the lines themselves. If these effects can be measured by an observer, the physical conditions can be inferred. This process is critical to the use of spectral lines to generate astrophysical understanding.

Zeeman Effect

The two electrons found in each orbital in an atom have different spins and hence different quantum states. If an external magnetic field is applied, the energies of each of these states shift differently, the degeneracy of the state is lost, and the transitions between energy states split. This is called the *Zeeman effect*,[21] and allows the magnetic field strength to be inferred.

In most astrophysical situations, the weak-field case will be the relevant one, in which the energy shift of the levels (and hence the frequency shift of the transitions) is proportional to the component of the external magnetic field *along the line of sight*, usually written B_{los} or $B_{\parallel}$. The line split itself will often be smaller than the linewidth, making it hard to detect; but the different components of a Zeeman-split line have different polarization characteristics (most visible as circular polarization). By observing the polarization of the spectral line, the Zeeman effect can be measured more accurately.

The Zeeman effect is most often used with OH transitions, but it has also been observed in the 21 cm atomic hydrogen line, and in water maser lines at 22 GHz.

External electric fields also cause energy level changes (the *Stark effect*), but this is generally only seen as line broadening (see below) in astrophysical systems.

Spectral Line Broadening: Local Effects

Spectral lines from any source—atomic or molecular—are never at a single frequency, but are instead found over a range of frequencies, with a distribution that is usually strongly peaked near the frequency of the transition. The simplest parameter to describe the frequency distribution is the linewidth, often given as the full width at half maximum power (FWHM) or half-power width. There are several effects that contribute to broadening the spectral line's frequency distribution.

Natural Broadening: One form of the *Heisenberg uncertainty principle*[22] is $\Delta E \Delta t \geqslant \hbar/2$, which implies that, for a quantum system in an excited state, we

---

[21] Pieter Zeeman (1865–1943), made this discovery in 1896 and was awarded the Nobel prize in 1902 (shared with Lorentz).
[22] Werner Heisenberg (1901–1976), German physicist. One of the founders of modern quantum theory.

cannot precisely know both the duration of the excited state and the energy of the state. This uncertainty in energy leads to an uncertainty in the frequency of the spectral line, and hence a distribution of line frequencies. The line profile due to natural broadening is described by the *Lorentzian distribution*[23] (see Appendix E).

Thermal Broadening: As atoms are always in motion due to thermal energy, the Doppler shift causes a broadening of the spectral lines. As the atoms move toward or away from the observer, the photons they absorb or emit are Doppler shifted from the normal spectral line frequency. Because the velocities associated with thermal energies are not relativistic, the non-relativistic Doppler formula can be combined with the Maxwell–Boltzmann thermal velocity distribution to find:

$$\Delta_\nu = \frac{2\nu_0}{c}\sqrt{2\ln 2\frac{kT}{m_0}} \qquad \Delta\lambda = 2\lambda_0\sqrt{2\ln 2\frac{kT}{m_0}}. \qquad (6.21)$$

This broadening of the spectral line has a Gaussian shape and the full width at half maximum is given by:

$$\Delta\lambda_{1/2} = 7.16 \times 10^{-7}\lambda\sqrt{\frac{T}{m_0}} \qquad (6.22)$$

with λ in Å, T in kelvin and m_0 in atomic mass units.

Pressure Broadening: Pressure broadening is caused mainly by collisions between the emitting particles and other particles in the environment (collisional pressure broadening). If the medium has a high enough pressure that the timescale between collisions is less than the lifetime of the excited state, the Δt in the Heisenberg inequality above will be defined by the pressure and not by the excited state lifetime. This will lead to a broader Lorentzian line profile.

Other forms of pressure broadening arise due to the influence of nearby particles on the emitting particle, giving rise to small shifts in the energy levels. (Since there must be nearby particles, this can only occur at relatively high pressure.) These shifts can occur through:

1. *Stark broadening*, where an external electric field due to a charged particle shifts the energy levels.
2. *Van der Waals broadening* caused by the Van der Waals[24] forces acting between particles. The spectral line shape is described by the Van der Waals distribution (see Appendix E).
3. *Resonance broadening*, where the interacting atoms are of the same element, allowing energy exchange between them.

Line Broadening Due to Bulk Motion: Line broadening through the Doppler effect due to thermal motion has been mentioned above, but an observed spectral line may arise from a large volume of gas with internal velocity structure. Since different parts

---

[23] Hendrik Lorentz (1822–1893), Dutch physicist, Nobel prize, 1902 (shared with Zeeman). Also created the Lorentz transformations, preceding relativity theory.
[24] Johannes Diderik van der Waals (1837–1923).

of the volume will be moving at different velocities, the spectral line will be a combination of shifted components.

One form of bulk motion that can broaden a line is turbulence in the emitting medium, in which parcels of gas are moving around randomly. In the case of molecular transitions arising from dense clouds, the turbulent broadening of the line is the dominant component, since the thermal linewidth is low in the cold gas.

If the spectral line samples rotating gas, broadening will also be seen, as part of the gas comes toward the observer, and part recedes. A 21-cm spectral line of atomic hydrogen that arises from an entire rotating galaxy is an example, and this can be used to estimate the galactic parameters: The more massive the galaxy, the faster its gas and dust must revolve about its center. This estimated mass can be converted to an estimated luminosity, which can in turn be compared to the brightness of the galaxy to yield a distance estimate (the *Tully–Fisher method*,[25] used only for spiral galaxies). A similar method, using the integrated stellar spectrum, can be used for elliptical galaxies (the *Faber–Jackson method*[26]).

Spectral Line Broadening: Transport Effects
Material between the source and observer can absorb some of the photons emitted by the source. This will generally only occur when there is intervening material of the same kind, which can absorb at the same frequency as the emission. The same kinds of frequency-broadening and -shifting effects as are found in the emitting material can be found in the intervening material, leading to complex lineshapes defined by both emission and absorption.

A fairly common situation is that a broad line may pass through absorbing material that can only absorb close to the transition frequency, so that absorption happens more in the central peak of the line than in the line wings. This will cause a decrease in the height of the emission line, and the full width half maximum will be measured to be higher. In extreme cases, the absorption may cause a dip in the center of the line, leading to a double-peaked line profile.

6.3 Further Reading

The treatment of molecular spectral lines in this chapter is quite cursory. These spectral lines are one of our most important ways to estimate the physical conditions of interstellar and circumstellar material. A reader wishing to undertake this analysis is advised to start with the treatment in *Tools of Radio Astronomy* (Wilson et al. 2013).

References

Blain, A. W., Smail, I., Ivison, R. J., Kneib, J.-P., & Frayer, D. T. 2002, PhR, 369, 111
Botteon, A. 2019, PhD thesis, Universita di Bologna
Burton, M. G. 1992, AuJPh, 45, 463
Draine, B. T., & Lee, H. M. 1984, ApJ, 285, 89

---

[25] Richard Brent Tully (1943–) and James Richard Fisher (1943–).
[26] Sandra Faber (1944–) and Robert Jackson (1949–).

Draine, B. T., & Li, A. 2001, ApJ, 551, 807

Hildebrand, R. H. 1983, QJRAS, 24, 267

Hughes, D. H., Dunlop, J. S., & Rawlings, S. 1997, MNRAS, 289, 766

Miyoshi, M., Moran, J., Herrnstein, J., et al. 1995, Natur, 373, 127

Omont, A., Cox, P., Bertoldi, F., et al. 2001, A&A, 374, 371

Reid, M. J., Dame, T. M., Menten, K. M., & Brunthaler, A. 2016, ApJ, 823, 77

Wilson, T. L., Rohlfs, K., & Hüttemeister, S. 2013, Tools of Radio Astronomy (Berlin: Springer)

Chapter 7

Particle Astrophysics: Gamma Ray, Cosmic Ray, and Neutrino Astronomy

High-energy astrophysical phenomena abound in the universe allowing us to study environments that could never be replicated on Earth. We examine in this chapter the history and detection of these high-energy messengers including gamma rays, cosmic rays, and neutrinos.

7.1 Historical Introduction

Particle astrophysics (or astroparticle physics) is a rapidly growing field that draws on theoretical insights and experimental methodology from high-energy particle physics to explore extreme astrophysical phenomena, and conversely, using astronomical observations to explore the behavior of matter in energetic and dense environments. Particle astrophysics examines elementary particles of astronomical significance and their connection to astrophysics and cosmology. It is a relatively new research area that is emerging at the intersection of particle physics and astrophysics, including cosmology. With cosmic rays reaching significantly higher energies than terrestrial experiments, particle physicists are looking to astronomy for novel experimental data. The cosmic ray spectrum, for example, includes particles with energies as high as 10^{20} eV, whereas the *Large Hadron Collider* (LHC; https://home.cern/) collides protons at an energy of 10^{12} eV.

This field has experienced rapid growth, both theoretically and experimentally, since the early 2000s, inspired by such astrophysical discoveries as neutrino oscillations (De Angelis & Pimenta 2018). Astroparticle physics historically focused on optical astronomy, before expanding as detector techniques improved. Many fields of physics, such as electrodynamics, thermodynamics, plasma physics, nuclear physics, special and general theory of relativity, and elementary particle physics are relevant to this research area.

The field is said to have been initiated in 1910 when, at the bottom and top of the Eiffel Tower, a German physicist named Theodor Wulf (1868–1946) measured the presence of ionized particles in the air, a sign of gamma radiation. He found that there was far more ionization at the top of the tower than would be expected from terrestrial sources alone. The Austrian physicist Victor Francis Hess (1883–1964) speculated that radiation from the sky was responsible for some of the ionization. To test his hypothesis, Hess developed instruments capable of making measurements at high altitude, up to 5.3 km. He made ten balloon flights from 1911 to 1913 to measure ionization levels. While his observations initially indicated that ionization levels decrease with altitude, he found that they rise significantly above a certain altitude. He realized that the ionization levels were much higher at the extremes of his flights than on the surface.

Hess inferred that radiation of very high penetrating ability enters our atmosphere from space. Two years later (in 1914), Werner Kohlhörster (1887–1946) confirmed this extraterrestrial radiation. Charles Thomson Rees Wilson (1869–1959) made it possible to identify and trace the tracks left by ionizing particles by creating the cloud chamber in 1912. After World War I, a renewed interest in radiation phenomena was sparked by a book by Walther Hermann Nernst (1864–1941) in 1921, which proposed that the penetrating rays originate from dying stars (Nernst 1921).

In this chapter, we will survey the most important topics of modern astroparticle physics:

- Nuclear astrophysics
- Dark matter
- Gamma-ray astronomy
- High and low-energy neutrino astrophysics
- Gravitational waves
- Cosmic rays.

7.2 Gamma-ray Astronomy

Astrophysical object observations in the optical spectrum is the purview of traditional astronomy. However, the optical spectrum encompasses just a minute portion of the overall electromagnetic spectrum. All aspects of this spectrum are used for astronomical observations as shown in Figure 1.1. From the longest wavelengths of radio astronomy to the sub-optical wavelengths of infrared astronomy, to optical astronomy, to ultraviolet astronomy, to X-ray astronomy, one eventually arrives at the shortest wavelengths of high-energy gamma-ray astronomy.

Gamma rays are the most energetic type of electromagnetic radiation. While there is no physical criterion that definitively separates gamma rays from X-rays, the latter generally have energies of 100 eV to 100 keV. X-rays are associated with emission from the innermost orbits of atoms, and temperatures of 10^6–10^9 K. When they encounter any matter, they have enough energy to ionize multiple atoms and break molecular bonds.

Gamma-ray photons have energies above 100 keV, which is enough energy for some very novel physical processes to be possible. The rest-mass energy of an electron is 511 keV, so two gamma rays with energy are able to create an electron–positron pair. Above a few MeV, a gamma-ray photon has energy roughly equal to the binding energy per nucleon of protons and neutrons inside nuclei. Above 500 MeV, a gamma-ray photon has enough energy to completely disintegrate an iron nucleus into its component protons and neutrons. In short, gamma-ray photons are able to participate in nuclear and particle physics processes that are more than mere scattering. The maximum energy of any observed photon is currently ~450 TeV (4.5×10^{14} eV).

In a landmark paper, Morrison (1958) presented the first estimates of fluxes from interstellar gamma-ray sources, produced by interactions of cosmic rays. In the 1960s, space-based observations by gamma-ray satellites such as Explorer XI were able to study photons with energy ~100 MeV (Kraushaar & Clark 1962). The field expanded rapidly with the first observation of gamma-ray bursts (GRBs) by Vela satellites (Klebesadel et al. 1973), and the first detection of gamma rays by Orbital Space Observatory 3 (OSO-3) through association of cosmic rays with interstellar Milky Way material (Kraushaar et al. 1972).

Observing gamma rays from the ground is more difficult. Because gamma rays react with the atmosphere, the original photon is lost, but secondary products are formed that can be observed on the ground. Careful observation and analysis can reconstruct the characteristics of the original astrophysical gamma-ray source. This remarkable technique was developed and improved primarily by groups in the US and France, such that the first astrophysical source of gamma rays (the Crab Nebula) could be identified from the ground at TeV energies (Weekes et al. 1989). To accurately capture the shape of the atmospheric shower of particles caused by the original gamma-ray photon and further remove the background, the key technological advance was a pixelated camera. Present-day instruments—including the High Energy Stereoscopic System Project (HESS), Major Atmospheric Gamma Imaging Čerenkov Telescopes (MAGIC), The High Altitude Water Čerenkov Experiment and High Altitude Water Čerenkov Observatory (also known as HAWC), Very Energetic Radiation Imaging Telescope Array System (VERITAS)—are fitted with pixelated cameras with more than one telescope covering a field of view of ~5° diameter and ~1000–2000 pixels in each camera (Hinton & Hofmann 2009). Our current view of the gamma-ray universe at different energies is shown in Figures 7.1 and 7.2.

7.3 Production of Gamma rays

Gamma-ray sources are generally the result of non-thermal emission mechanisms. Unlike thermal radiation arising from the random motions of particles, non-thermal radiation can have a number of origins: de-excitation or decay of atomic nuclei, annihilation of particles with their antiparticles, and the interaction of photons and matter between non-thermal particle species. These processes will either contribute to mono-energetic photon emissions or to photon continuum emission that spans the electromagnetic spectrum from the radio band to gamma rays. Radioactive decay

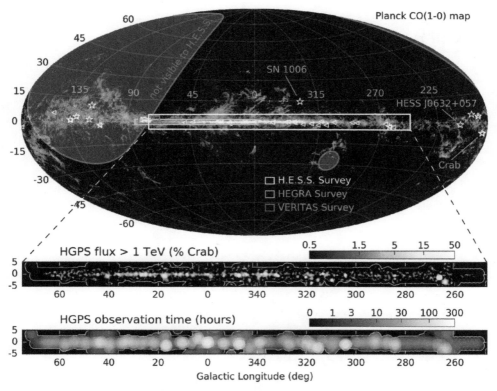

Figure 7.1. The HESS Galactic plane survey (HGPS). Image credit: H.E.S.S. Collaboration et al. (2018) reproduced with permission © ESO.

and particle-antiparticle annihilation processes can produce copious amounts of gamma radiation. PWN, neutron stars and pulsars, SNRs and micro-quasars are plausible astrophysical sites for such processes. Outside the Milky Way, gamma rays have been detected from galaxies with very high rates of star formation and ultra-relativistic jets of particles from AGN surrounding super massive black holes (SMBHs).

Gamma rays can also be produced by relativistic particles in strong magnetic fields. When accelerated charged particles such as nuclei, electrons or positrons interact with the interstellar medium, magnetic or radiation fields, gamma rays are produced. These targets for radiation are established through multiwave band studies, i.e., pion decay and bremsstrahlung (Section 6.1.1). Synchrotron emission and inverse Compton scattering depend on, and thus can be used to constrain, the energy distribution of energetic charged particles (Figure 7.3). The examination of cosmic gamma rays indicates fascinating possibilities relevant to the indirect quest for dark matter (DM; Bergström & Snellman 1988) and the discovery of new basic laws of physics, such as the violation of Lorentz invariance (Amelino-Camelia et al. 1998).

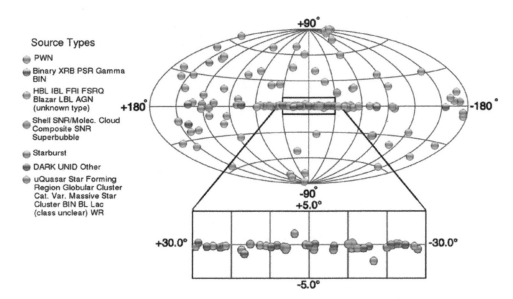

Figure 7.2. VHE gamma-ray sources on the sky in galactic coordinates. Image credit: Reprinted from Degrange & Fontaine (2015), Copyright (2015), with permission from Elsevier.

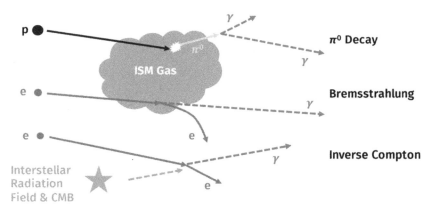

Figure 7.3. Common mechanisms of gamma-ray production. Image credit: Sabrina Einecke.

Bremsstrahlung and synchrotron radiation are discussed in Chapter 6 (Sections 6.1.1 and 6.1.2). Here, we discuss other forms of production relevant to gamma rays.

7.3.1 Inverse Compton Scattering

Inverse Compton Scattering occurs when a low-energy photon collides with a high-energy charged particle, usually an electron. For a sufficiently energetic (relativistic) electron, the photon will emerge with gamma-ray energies. Thanks to the cosmic microwave background, a ready supply of low-energy photons is always on hand. As a result, sufficiently energetic electrons are a significant astrophysical source of gamma rays.

7.3.2 π^0 Decay

Proton–proton or proton–nucleus interactions can produce charged and neutral *pions*, which are subatomic particles that consist of a quark and an anti-quark ($u\bar{d}$, $d\bar{d}$, $u\bar{u}$, or $d\bar{u}$). As shown in Figure 7.4, charged pions decay into muons and neutrinos with a lifespan of 26 ns, while neutral pions decay within 8.4×10^{-17} seconds into two gamma-ray photons as follows:

$$\pi^0 \rightarrow \gamma + \gamma. \tag{7.1}$$

Because pions are most efficiently generated at (relatively) low energies, gamma-ray photons from pion decay have energies of usually 70 MeV.

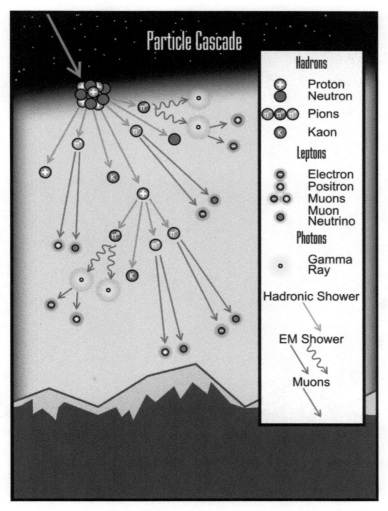

Figure 7.4. Particle cascade and π^0 decay. Image credit: Theturnipmaster/Wikipedia.

7.3.3 Annihilation of Matter–Antimatter

Just as particle pairs can be created by energetic photons (pair production), particles can annihilate with their antiparticles to produce photons. Electron–positron ($e^+ + e^- \rightarrow \gamma + \gamma$) and proton–antiproton annihilation (see Figure 7.5) are the most common instances of this mechanism.

If photons are produced, conservation of momentum implies that at least two photons must be produced. For electron–positron annihilation, the photons obtain 511 keV energy, each equivalent to the rest mass of the electron or positron. Proton–antiproton annihilation is more complicated because the proton is a composite particle. For example, the interaction may produce pions, which then decay into photons:

$$p + \bar{p} \rightarrow \pi^+ + \pi^- + \pi^0. \tag{7.2}$$

7.3.4 Fermi Acceleration

Fermi[1] acceleration involves the continual acceleration of charged particle by repeated interaction with a shock front. As originally formulated, particles would be continually accelerated by head-on collisions with magnetized interstellar clouds (acting as magnetic mirrors). This known as second-order Fermi acceleration (Figure 7.6, right). A similar but more efficient mechanism was later discovered,

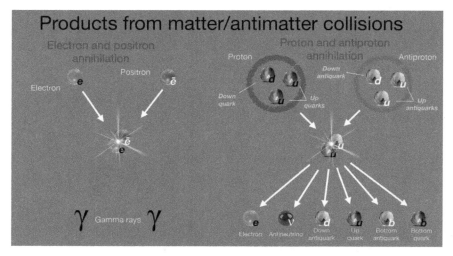

Figure 7.5. Matter–antimatter collisions create a variety of products, depending on the incident particles. When electrons and positrons annihilate, they make gamma rays. Protons are composed of quarks (and antiprotons of antiquarks); these collisions create more complicated particle interactions.

---

[1] Enrico Fermi (1901–1954), an Italian physicist and the "architect of the nuclear age, the atomic bomb and famous Fermi paradox."

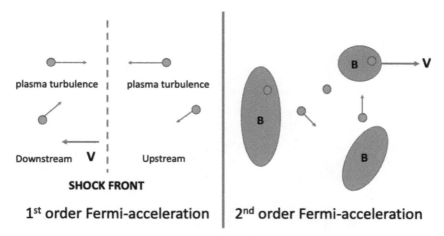

Figure 7.6. (Left) First-order Fermi acceleration. Particles are accelerated by shock waves. (Right) Second-order Fermi acceleration, where cosmic rays are accelerated by random moving magnetic clouds.

in which particles are repeatedly accelerated at a shock front. This known as first-order Fermi acceleration (Figure 7.6, left), also referred to as diffusive shock acceleration.

Fermi acceleration, particularly of the 1st-order kind, is believed to be the primary process by which charged particles obtain non-thermal energies through astrophysical shocks. Shocks are very common in astrophysical environments, including supernova explosions, galaxy–galaxy collisions, collisions of Earth's magnetic field with solar flares, stellar and galactic jets, and accretion flows. In all situations, in order for the process to be successful, the system must be effectively collisionless. Inter-particle collisions would quickly establish a thermal spectrum of particle energies, bringing ultra high energy particles back to average.

Typically, shock waves travel through both up- and downstream magnetic inhomogeneities. Consider a charged particle passing through a shock front. The shock wave will accelerate the particle as it passes the front. On average, most particles will join the hot, post-shock material downstream from the shock front. But because of thermal velocity dispersion, turbulence, and/or dispersion along tangled magnetic field lines, a fraction of particles may return to the upstream region and be accelerated again. Multiple interactions increase the energy considerably. This mechanism of acceleration of charged particles is first-order Fermi acceleration. The resulting energy spectrum of particles follows a power law:

$$\frac{dN}{d\varepsilon} \propto \varepsilon^{-\alpha}, \tag{7.3}$$

where α is the *spectral index* and it depends only on the compression ratio of the shock.

Second-order Fermi acceleration involves acceleration due to repeated interaction of charged particles with randomly moving magnetic mirrors (see Figure 7.6, right). Particles will be accelerated by head-on collisions, and decelerated by receding collisions. As noted by Fermi in 1949, head-on collisions are more likely to occur, and so this mechanism can accelerate particles to high energies. Fermi considered this mechanism as a possible cause of cosmic rays. Given a modern understanding of the interstellar medium, this process is less efficient than first-order acceleration.

7.3.5 Electric Field Acceleration

Acceleration by electric fields is well known from elementary physics. However, special conditions are required in astrophysical systems to produce a significant electric field, given that the high conductivity of plasmas will usually allow efficient short-circuiting of the field through the motion of charged particles. *Unipolar inductors*, like spinning black holes or neutron stars, or local regions where opposite orientations of magnetic field combine are examples of astrophysical environments in which electric fields play a part. Where electric and magnetic fields are parallel, magnetic field reconnection can very efficiently accelerate charged particles. The electron–electron or proton–proton mean free path for collision is given in Petrosian & Bykov (2008):

$$\lambda_{\text{collision}} \sim 15 \text{ kpc} \left(\frac{T}{10^8 \text{ K}} \right)^2 \left(\frac{10^{-3} \text{ cm}^{-3}}{n} \right) \tag{7.4}$$

where n is electron or proton density and T is temperature. The energy gained by acceleration is smaller than the collision losses and only a limited percentage of the particles can be accelerated to a non-thermal energy range.

7.4 Detection of Astrophysical Gamma Rays

Detecting gamma rays is not easy. The number of gamma-ray photons we receive from a celestial source is very low and thus most sources require an integration time of at least few days to gather enough gamma-ray events to be certain of the detection. During that integration time, the number of false triggers can be enormous.

Because gamma-ray photons are highly energetic, they have small interaction cross-sections and so tend to ignore most materials in their path. There is no known way to focus gamma-ray photons onto a detector. Therefore, the aperture of the telescope is the aperture of the detector. Increasing the detector area will detect more gamma rays, but also increases the number of false triggers and of course, the cost.

The detection of a gamma ray relies on its interaction with matter. One of three events can affect the gamma-ray photon as it passes through matter: the Compton effect, photoionization or electron–positron pair production. Of these, photoionization is the most useful for detection because it is the only process which absorbs the entire photonic energy. The other two simply scatter the photon, absorbing only part of its energy. It is possible for the scattered photon to escape the detector, making it impossible to infer its original energy. Any of these events create motion of an electron, which can be viewed as a small electric current. This current can be amplified and measured. Gamma-ray detectors can be broken into two classes: *spectrometers* and *imaging detectors*.

The atmosphere prevents many gamma rays from reaching the Earth. Direct identification thus requires space-based instruments. As shown in Chapter 1, early gamma-ray telescopes were balloon based. The detector was mounted in the balloon payload. The system is stabilized by a three-gyroscope platform system and pointing is accomplished with the aid of known stars. Balloon-based observations work well for strong gamma-ray sources like the Sun, but weaker sources need longer integration times. Even then, it was obvious that these sources must be observed from space-based platforms The detection area for space-based detectors is inadequate at higher energies, so an indirect ground-based technique is also required.

Ground-based detectors work on the principle of detecting the interaction of gamma rays with the atmosphere and producing secondary products. A particle shower in the atmosphere is formed in this reaction, consisting mainly of electrons and positrons (gamma-ray initiated shower) or electrons and muons (proton initiated shower). For example, the *Imaging Atmospheric Čerenkov Telescopes (IACTs)* detect the Čerenkov light produced by atmospheric shower particles, and *Čerenkov Water Detectors* detect the particles directly by the Čerenkov light generated by the ground water detectors. Since gamma rays must reach the ground in the case of water detectors, their threshold is typically higher (~100 GeV), but they can operate continuously while the *IACTs* are limited to operating at night so that they can detect faint Čerenkov light.

7.4.1 Ground-based Detectors

In recent times, *Imaging Atmospheric Čerenkov Telescope* (IACT) devices have become the most sensitive detectors for the detection of gamma rays beyond ~50 GeV. In the focal plane of the mirrors, they map the Čerenkov light of the air showers with mirrors to a fast camera and thus capture the angular distribution of the Čerenkov light from the air shower. The detector used to monitor Čerenkov radiation (Section 6.1.3) generally consists of a large, segmented mirror reflecting the Čerenkov light on a series of photo-multiplier tubes. High-speed electronics digitize, amplify, and register the image of the shower. A collection of such telescopes, which can be placed 70–120 m apart, is the most powerful mode of operation. Closer spacing helps low-energy efficiency to be increased at the cost of the collection area at higher energies. The total amount of light captured is proportional to the total length of the track of the particle shower, which, in essence, is proportional to the energy of the primary particle, giving a calorimetric calculation of the shower.

Three current observatories that use the technique of stereoscopic *IACTs* are:
- *HESS* is a network of four IACTs originally found in the hills of Khomas in Namibia. The initial four HESS telescopes have a 107 m^2 mirror area with a 1 m focal length. At the trigger stage, the energy threshold at the zenith is ~100 GeV.
- *MAGIC (Major Atmospheric Gamma-ray Imaging Čerenkov Telescopes)* system consists of two telescopes situated at an height of 2400 m on the Island of La Palma. The two telescopes have a huge 240 m^2 mirror area,

corresponding to the ~50 GeV energy threshold. They are designed in a light-weight manner to allow quick slewing to take place and able to follow up immediately on the signals of gamma-ray bursts (GRBs).

- *VERITAS (Very Energetic Radiation Imaging Telescope Array System)* is a system of 4 IACTs with a mirror size of 106 m^2 that is similar in sensitivity and efficiency to *HESS*, with a smaller field of view.

Water Čerenkov detectors work at a higher threshold of energy than IACTs since they measure particles of the shower that reach the ground. These detectors are usually located at higher altitudes for this purpose. The energy that reaches the ground is about 10% of the initial energy. The detectors have characteristics that allow a comprehensive study of the gamma-ray sky at TeV energy ranges. They are well suited to researching transient sources and conducting sky surveys by incorporating a wide field of view and almost a 100% duty cycle. Water Čerenkov detectors work under the theory that the Čerenkov light in the water tanks is observed from the shower particles. To reconstruct the energy of the incident photon, the detector serves as a calorimeter. The HAWC (High Altitude Water Cherenkov Observatory, located in Mexico) is the most prominent example of a gamma-ray and cosmic ray observatory.

7.4.2 Space-based Detectors

Space-based detectors over 20 MeV work via pair-creation in the detector. The *Fermi Gamma-ray Space Telescope (FGST)*, launched in June 2008, is the primary instrument in this energy spectrum. The *Fermi-LAT (Large Area Telescope)* produces measurements between 20 MeV and 300 GeV in the energy spectrum. The system consists of a tracker to measure the traces of the electron–positron pair produced in a pair of gamma rays, a calorimeter to evaluate the energy of this electron–positron pair, and then a primary gamma ray and an *anti-coincidence detector* to track the background charged particles. The recent status of ground and space-based instrumentation with corresponding flux sensitivity and energy range is shown in Figure 7.7.

7.4.3 Gamma-ray Spectrometers

Gamma-ray spectrometers collect gamma rays by scintillation. Scintillation transforms the gamma-ray photon into a visible light photons, which are then detected with a photo-multiplier tube (PMT). There are several types of Gamma-ray spectrometer.

Crystal Scintillators
The scintillation material is generally made from an alkali-halide salt, such as sodium iodide (NaI) or cesium iodide (CsI). To aid in the scintillation process, these crystals contain a small amount of impurity, such as thallium or sodium; these detectors are often listed as NaI(Tl) or CsI(Na). The impurity is called the *activator*.

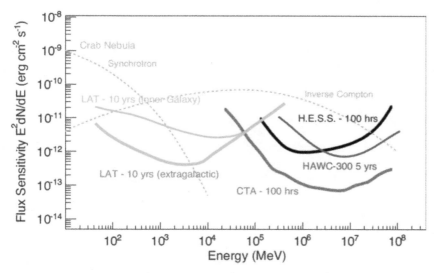

Figure 7.7. Status of instrumentation with flux sensitivity and energy. The "inner Galaxy curve" refers to the approximate background at position $l = 10°$; $b = 0°$, while the "extragalactic curve" is measured at position $l = 10°$; $b = 90°$. Image credit: Funk et al. (2013), Copyright Elsevier (2013). CC BY 3.0.

NaI(Tl) crystals are the most commonly used (inorganic) scintillators because they have the greatest light output per event.

Another common material is bismuth germanate (BGO), $Bi_4Ge_3O_{12}$, which is a cubic lattice crystal. It produces thousands of photons in the visible band when struck by a gamma ray. The number of photons produced depends on the energy of the gamma-ray photon. These photons are sensed by a photo-multiplier tube.

There are also organic scintillators, including plastics such as polyvinyl toluene or polystyrene. Many other organic substances, including liquids, also exhibit scintillation. Organic scintillators rely on electron transitions in the carbon-based molecules to detect the absorption of a gamma ray.

The probability that a gamma ray entering the detector produces a count is a measure of the detector's efficiency. Generally, larger volume detectors are more efficient than smaller, but the detector shielding plays an important role in that relationship. Higher efficiency detectors can produce spectra faster than low because they have a shorter "dead time" during which data from an event is collected. The efficiency can be stated in absolute or in relative terms. Absolute efficiency values are given in percent, and describe the probability that a gamma ray of a particular energy will interact with the detector. Detector efficiency is measured by taking a spectrum from a known source and comparing the count rates in each peak to the expected count rates for the known source. By plotting the detector efficiency at various gamma-ray energies, an efficiency curve is created. The curve is then used to determine the detector efficiency at other gamma-ray energies.

There are two ways of describing the resolving power of a scintillator detector. The most common is the full width at half maximum (FWHM). This is the width of the beam at half the maximum value of the peak. The resolution is specified at the energy level being observed. If for example, the resolution is given as FWHM = 7.5 keV at 120 keV, then the resolution is specified in absolute terms A relative resolution can be given by stating the percentage of the two values, 7.5/120 = 6.25% giving a relative resolution of 6.25% at 120 keV.

Semiconductor Detectors

Semiconductor detectors, also called solid-state detectors, operate very differently than scintillation detectors. An electric field is placed across the volume of semiconductor crystal material. When a gamma ray enters the crystal, it kicks an electron from its valence band energy into the crystal's conduction band. This creates an electron–hole pair in the semiconductor. The electric field then causes the electron to move toward the positive electrical connection and the hole to move toward the negative connection, establishing a small current. This current is sensed and amplified, with the signal sent to the signal processing and data analysis system.

Typical semiconductor detectors are made from germanium (Ge), cadmium telluride (CdTe) or cadmium zinc telluride (CdZnTe). Relative efficiency for semiconductor detectors is measured by comparing the detector against a NaI crystal scintillator. This can lead to stated efficiencies of semiconductor detectors of greater than 100%.[2] Germanium detectors provide the highest efficiency, with resolutions of 500 or greater. Relative energy resolution is typically less than 1%.

Gathering the Data

An interaction between the gamma ray and the detector crystal creates an event that is transformed into an electrical signal, generally a small voltage spike. This voltage is proportional to the energy of the interaction, which in turn is proportional to the energy of the gamma-ray photon. The voltage spike is passed to an analog-to-digital converter (ADC) so that the generated numerical value represents the magnitude of the voltage spike. The ADC then sorts these numbers into "bins" or "channels." The number of channels is always a power of two, and depends on the resolution of the system and the energy range being detected. The output of the ADC is then sent to a multichannel analyzer (MCA) where the channel distribution is reshaped into a Gaussian shape. This is sent to a computer for recording and processing.

If the data (counts) in the channels of the MCA are plotted, there will be peaks lying at one or more channels. These peaks are called spectral lines, by analogy with optical spectra. The x-axis position of the peak is related to the gamma ray's energy and the area of the peak is related to the intensity of the gamma-ray radiation and the efficiency of the detector. The width of the peak is related to the detector's resolving power.

7.4.4 Gamma-ray Imaging Detectors

There are a few different methods for producing images of gamma-ray sources. Below, we discuss *Coded Mask*, *Crystals and Spark Chambers*, and *The Laue Lens*.

---

[2] The percent symbol is often neglected.

Coded Masks

Gamma-ray imaging detectors rely on either pair production or Compton scattering techniques working in conjunction with a coded mask to determine the source direction of the incoming gamma ray. The principles of the coded mask are based on the optics of the pin-hole camera. A single pin hole would allow imaging of the source, but the intensity is too low for this to be practical. To increase the sensitivity of the imaging system, the number of pin holes is increased. This then produces a number of source images equal to the number of pin holes, with the images overlapping, much like Aristotle seeing hundreds of Sun images through the shadow pattern of tree leaves. If the hole pattern is known (it generally matches the detector's pixel array), a faithful image can be reconstructed.

Research in the late 1960s and early 1970s showed that the idea worked, but the hole pattern had to be chosen carefully or false images (false sources) are created. The research produced a class of patterns called *uniformly redundant arrays* or coded masks. Using these patterns, the shadow pattern of any gamma-ray source in the image is not affected by the sources at any other position in the image. The mask is larger than the detector and the shadow of the mask from any one source is positioned on the detector in a different position than the shadow of the mask from any other source.

To reconstruct the image, the entire detector signal is cross-matched with each possible source mask position. Where there is a correlation, the count is increased and where there is not, the count is decreased. With the mask, each source exposes about half the detector's pixels. If there is no source, the counts sum to zero. The signal-to-noise ratio of the imager is given by the square-root of the number of holes in the mask. Unfortunately, if the gamma-ray source distribution is such that the imager aperture is uniformly illuminated, the mask acts as a single pin hole, and the source or sources cannot be imaged.

Crystals and Spark Chambers

Another imaging technique uses an NaI crystal above a gas-filled spark chamber. Here, the gamma ray interacts with the crystal to produce an electron–positron pair which pass through the chamber, ionizing the gas. The chamber has layers of criss-crossed wire electrodes. The gas ions (and freed electrons) are picked up by the electrodes and in this way, the path of the particles through the gas chamber can be determined. The pair is captured by another detector at the bottom of the chamber so their energies can also be measured. With knowledge of the particle pair paths and energies, the direction of the incoming gamma ray can be determined. This technique does not require a coded mask.

The Laue Lens

The Laue[3] lens uses Bragg diffraction in a pattern of mosaic crystals (in this case, made of copper) to bring the gamma rays and X-rays to a focus point behind the lens

---

[3] Max von Laue (1879–1960), German physicist, Nobel prize 1914, for his discovery of the diffraction of X-rays by crystals.

structure, on the surface of a position sensitive detector. This diffraction was first proposed in 1913 by W. L. Bragg[4] and W. H. Bragg. The theory was created to explain their discovery of patterns in reflected X-rays from crystalline solids. They found intense peaks of reflected radiation (known as Bragg peaks) for certain specific wavelengths and incident angles. W. L. Bragg explained the result by modeling the crystal as a set of discrete parallel planes with constant spacing. The incident X-ray radiation produces a Bragg peak if the reflections off the various planes interfere constructively.

Mosaic crystals are those containing microscopically perfect crystals with their lattice planes slightly misaligned from each other, around an average direction. The alignments are described by an (approximately) Gaussian distribution function. The FWHM of the distribution defines the mosaic spread of the crystal. Given this mosaic property, the lens is able to focus more than a single wavelength of photons. The wider the mosaic, the greater the lens' bandwidth.

7.5 Astrophysical Gamma-ray Sources

Of all Galactic sources of gamma rays, the primary contribution (nearly 80% of the GeV energy range) to the gamma-ray flux is related to the *Galactic diffuse emission* under TeV energies, first detected by OSO-3, and then confirmed by SAS-2 as well. Figure 7.8 displays this diffuse emission, measured by Fermi-LAT. A significant fraction of this emission above 30 MeV is probably due to:

1. Cosmic rays interaction with interstellar gas producing *neutral pion decay*, which contributes to soft gamma-ray emission, and

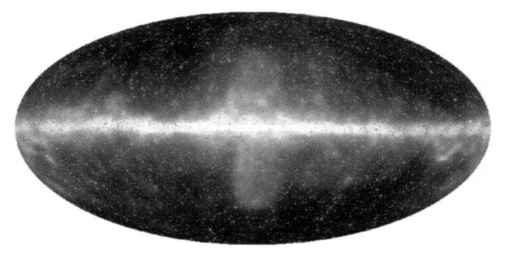

Figure 7.8. View of sky through gamma rays above 600 MeV energy, measured by Fermi-LAT. Red ~1 GeV and Blue ~300 GeV. Image credit: Selig et al. (2015) Copyright EDP Sciences.

---

[4] Sir William Lawrence Bragg (1890–1971), Australian physicist, Nobel Prize 1915, shared with his father Sir William Henry Bragg (1862–1942), English physicist and chemist.

2. Cosmic ray electrons scattering through *interstellar radiation fields*, contributing to the hard gamma-ray emission. This *Galactic diffuse emission* was also detected in TeV energies by HESS in the innermost region of the Galaxy (Abdalla et al. 2018) and also by HAWC in a large-scale zone of the Galactic Plane (Abeysekara et al. 2017).

Two large bubble-like structures, spreading $\sim 50°$ below and above the Galactic Center, have been detected in addition to the diffuse emission. The gamma-ray emission from these systems, called the *Fermi bubbles*, shows a power-law spectrum, which is considerably harder than the galactic disk diffuse emission. These bubbles can be seen in Figure 7.8. Acceleration of cosmic rays by the SMBH Sgr A* at the heart of our Galaxy, an event of starburst action, or a long-term buildup of cosmic rays from the regular formation of stars near the Galactic Center are some of potential causes for the formation of these bubbles (Crocker & Aharonian 2011).

The large number of individual emitters that have been detected over the past decade is among the most remarkable findings in the gamma-ray sky. To date, over ~ 5000 sources above 100 MeV have been detected by the 8 yr Fermi-LAT survey, as seen in Figure 7.9. More than 3000 sources were contained in the third *Fermi-LAT catalogue* (Acero et al. 2015), resulting from 4 years of data processing. Around 80% are associated with extragalactic objects, while a significant fraction of the remainder are pulsars. One of the best-studied sources of gamma rays in our Galaxy is the *Crab Nebula*. It's a typical *Pulsar Wind Nebula* (PWN), the residue of a supernova explosion that took place in 1054 A.D. The wind is driven by the pulsar PSR B0531+21. In Figure 7.10, we show its multiwavelength SED.

Two-hundred sources have also been observed at higher energies (~ 100 GeV) and are compiled in the *TeV Catalogue*. Nearly half of these objects are located in our Galaxy, including Supernova Remnant (SNR)s, pulsars, binaries, PWN, and a large fraction of unclassified Galactic sources. The other half of these gamma-ray sources are extragalactic. They are most likely starburst galaxies and AGN. The canonical model of AGN emission pictures a SMBH, with mass up to $\sim 10^9 M_\odot$, at the center of an AGN. Gravitational energy is converted into kinetic energy and then radiation energy as material falls into the SMBH. Some of this energy is converted into kinetic energy of an outflow, creating highly-collimated relativistic plasma jets. Extragalactic TeV sources are most often *blazars*: a type of AGN whose jet is pointed at the observer, so that we are "looking down the barrel."

Cosmic objects emitting brief outbursts of gamma rays were discovered by the US Vela satellites in the early 1970s. The intention of these satellites was to monitor the testing of nuclear warheads in the atmosphere. However, the gamma rays detected did not originate from the Earth's surface or the atmosphere, but instead from external sources. These phenomena are known as gamma-ray bursts, or GRBs. They are transient: *short GRBs*, lasting a fraction of a second, are associated with mergers of neutron stars with neutron stars (Bahcall 1989). *Long GRBs* last upwards of a few seconds, and are connected with the energetic supernova of a very massive star ($\sim 10^2 M_\odot$). They are also accompanied, after minutes, hours, or days, by afterglows. Their energy spectrum is non-thermal and variable, reaching a peak at around

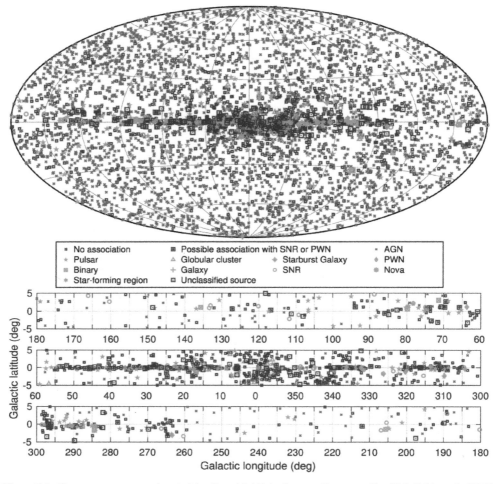

Figure 7.9. Gamma-ray sources detected by Fermi-LAT in 8 years. Image credit: Abdollahi et al. (2020) Copyright (2020) IOP Publishing.

several hundred keV and extending up to several GeV. A map of detected GRBs to date is shown in Figure 7.11.

In summary, the following are known astrophysical sources of gamma rays:

- Pulsar and pulsar wind nebulae (PWNe),
- Active galactic nuclei (AGN) and accreting black holes,
- Supernova remnants (SNRs),
- Unspecified VHE gamma-ray sources,
- Gamma-ray bursts (GRBs),
- Galactic diffuse emission,
- Stellar clusters and stellar winds,
- Sources of ultra high energy cosmic rays (UHECR),
- PeVatrons, and
- the Fermi bubbles.

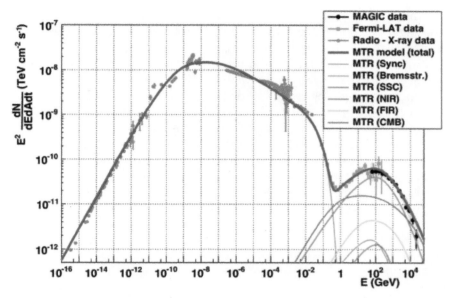

Figure 7.10. Spectral Energy Distribution for *Crab Nebula* in radio to gamma-rays range. Image credit: Yuan et al. (2011) Copyright (2011) IOP Publishing.

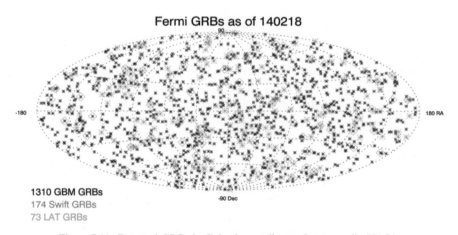

Figure 7.11. Detected GRBs in Galactic coordinates. Image credit: NASA.

7.6 Cosmic Ray Origins and Spectra

The term *cosmic ray* refers to electrons, nucleons, nuclei, gamma-ray radiation, and more exotic particles (such as muons and pions) of extraterrestrial origin. Cosmic rays reaching the Earth's upper atmosphere consist roughly of 70% protons, 15% alpha particles, 12% carbon, nitrogen, and oxygen nuclei, 1% electrons, and <0.1% gamma rays, with the remainder unidentified.

Particle accelerators propel charged particles (protons, electrons, ions) to relativistic speeds using intense, synchronized electric and magnetic fields. Cosmic particle accelerators are believed to do the same, but with perhaps different processes than machines built by humans. In the cosmos, the acceleration of particles could be a rapid, single-step process involving strong, rapidly-changing electric and magnetic fields, such as those surrounding rotating neutron stars or magnetars. Or, particle acceleration may involve incremental increases over many millennia. This process may occur in shock waves generated by supernova explosions, where the charged particles travel through twisted magnetic fields, gradually gaining energy until they make their high-energy escape.

High-energy gamma rays are almost always a non-thermal byproduct of cosmic particle accelerators. For example, a relativistic proton created by a supernova can collide with a heavy nucleus, creating a shower of new particles. Among them, the π^0 meson will quickly decay into two gamma rays. If there is a significant flux of high-energy electrons interacting with the material of an extended object, such as a SNR, the electrons can lose their energy through any of the interaction mechanisms described in the next chapter, producing non-thermal gamma rays.

Even if detectors on Earth can measure the incoming path of a cosmic ray, the charged particles will have had their path bent by Galactic magnetic fields and the magnetic fields of any objects encountered on the way. Therefore, it is near-impossible to determine their origin. On the other hand, gamma rays—produced by the interactions of cosmic rays with dense clouds or other objects—do travel in straight lines from their source, allowing us the possibility of studying the mechanisms creating cosmic rays. The interaction mechanisms producing gamma rays create a close relationship between the energy spectrum of the gamma rays and the energy spectrum of the accelerated particles producing the gamma rays.

The cosmic ray spectrum has been the subject of intense study. The spectrum measures the number of cosmic rays at various energies. Cosmic rays have been observed in the energy range from 10^9 eV to more than 10^{20} eV. Within this eleven decade range, the population of particles approximately follows a single power-law $\propto E^{-3}$. There are minor features at roughly 5×10^{15} eV and at 5×10^{18} eV, called the "knee" and "ankle" respectively (see Figure 7.12; there is evidence for a "second knee," which would physiologically entail that the cosmic ray spectrum belongs to an ostrich). Cosmic rays above 10^{19} eV are called ultra high-energy cosmic rays (UHE-CRs). At this energy level, the subatomic particles that comprise cosmic rays are carrying at least a joule of kinetic energy. The highest energy cosmic rays observed are carrying as much energy as a tennis ball traveling at 40 meters per second.

A variety of theoretical models have been proposed to explain the origin of knee and ankle features found in the spectrum. Current suggestions include that cosmic rays below 10^9 Gev are predominantly of Galactic origin, while above 10^9 Gev are of extragalactic origin. They proposed this different origin of cosmic rays as a possible explanation for the features observed in the cosmic ray spectrum.

7.7 Neutrino Astronomy

Neutrinos are light neutral particles that interact with matter very infrequently.[5] To understand how tiny they are, if a single proton were the size of the Earth, solar

---

[5] For a historical overview see Section 1.3.7.

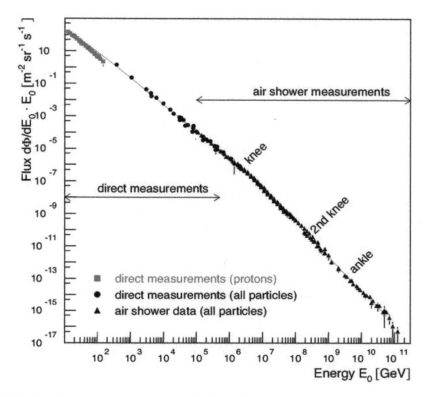

Figure 7.12. Cosmic ray spectrum. The spectral breaks known as knee (at $E \sim 3$ PeV), second knee (at $E \sim 400$ PeV), and ankle (at $E \sim 4$ EeV) are indicated. Image credit: Reprinted from Blümer et al. (2009), Copyright (2009), with permission from Elsevier.

neutrinos behave as if they were the size of a marble. It's definitely not easy to see neutrinos!

As they travel in space without absorption, deflection, or scattering, neutrinos are perfect astronomical messengers. However, their weak interaction makes them extremely hard to detect; large-scale detectors are required for neutrino observatories. Direct tests of the hypothesis of stellar evolution were provided by the observation of neutrinos released by the Sun and by nearby supernovae. These pointed to physics beyond the standard model, as they imply that neutrinos have a tiny but non-zero mass. Current and future neutrino detectors will track the most efficient sources of energy and particles in the Universe. The expectation is that the discovery of extragalactic neutrinos, most likely driven by mass accretion into black holes, may not only facilitate the study of the sources, but will also provide new knowledge on the basic properties of matter.

The weaker interaction of neutrinos implies that multiple kilotons of detecting material are required to maximize the chances of an interaction. Although the probability of a single neutrino interacting with the detector is very low, the large flux of neutrinos from the Sun—around 100 billion $cm^{-2}\,s^{-1}$—enables the identification of hundreds of interactions each year. However, extraneous particles, including neutrinos formed by cosmic ray collisions, create confusion in the detector. To prevent this, detectors are usually buried underground.

These detectors were first developed in order to provide direct experimental testing of stellar structure and evolution models (Bahcall 1989). While we believe that the Sun is fueled by the fusion of hydrogen into helium at 15 million K, we can only directly see the surface of the Sun, at a mere 5800 K. The photons produced by fusion—as four hydrogen atoms become one helium atom and the extra mass m is transformed to energy, in accordance with $E = mc^2$—can take hundreds of thousands of years to scatter from the core to the surface. How can we "see" into the core?

In the mid-1960s, it was proposed that we can test this model by looking for neutrinos expected to be released during fusion. In contrast to photons, the weak interaction of neutrinos with matter enable them to stream directly from the center of Sun and to Earth's detectors. Thanks to modern detectors, we can indeed "see" into the core of the Sun.

7.7.1 General Properties of Neutrinos and Their Discovery

As briefly introduced in Section 3.2.2, there are three types of neutrino, referred to as "*flavors*":

- *electron neutrino* (ν_e),
- *tau neutrino* (ν_τ), and
- *muon neutrino* (ν_μ).

These neutrons are associated with, and are produced in interactions involving, the corresponding leptons: the electron (e), muon (μ), and tau (τ). These particles all have charge of -1, but have widely-differing masses: the muon is about 200 times more massive than the electron, and the tau about 3500 times more massive than the electron. Each neutrino flavor has a corresponding *antineutrino*. The *lepton number* of each flavor—defined by adding one for each electron and each electron neutrino, and subtracting one for each positron and each anti-electron neutrino, and similarly for the muon and tau—is conserved in standard model interactions. Thus, we see interactions of the form:

$$\nu_q + A \rightarrow q^- + B \tag{7.5}$$

where $q = e, \mu, \tau$, and A and B are other reactants/products.

The most common interactions involving the electron neutrino and antineutrino are:

$$p + e^- \rightarrow n + \nu_e \tag{7.6}$$

$$n + e^+ \rightarrow p + e^- \tag{7.7}$$

$$p + \bar{\nu}_e \rightarrow n + e^+ \tag{7.8}$$

$$n + \nu_e \rightarrow p + e^- \tag{7.9}$$

$$n \rightarrow p + e^- + \bar{\nu}_e \tag{7.10}$$

$$p \rightarrow n + e^+ + \nu_e. \tag{7.11}$$

The above reactions are called β^- *capture*, β^+ *capture*, *inverse Beta decay (IBD)*, *IBD by neutron*, β^- *decay*, and β^+ *decay*, respectively. For energy conservation reasons, the final decay can only happen in the context of a nucleus; free protons do not beta decay.

The Austrian physicist Wolfgang Pauli (1900–1958) predicted electron neutrinos (ν_e) in 1930 to explain the significant energy lost in the process of radioactive beta decay. The Italian physicist Enrico Fermi further developed the theory of beta decay (1934) and gave this *ghost particle* the name neutrino, meaning "little neutral one" in Italian. An electron–neutrino is produced with a positron in β^+ decay, and an electron–antineutrino is released with an electron in β^-. Finally, the existence of the *electron–antineutrino* was confirmed in 1956 by a team of American physicists led by Frederick Reines (1918–1998). *Antineutrinos* produced in a nuclear reactor interacted with protons to yield neutrons and positrons. Evidence of the presence of the *electron–antineutrino* was provided by the peculiar energy signatures of the by-products.

The discovery of a second generation of charged lepton, namely the muon (μ^-), lead to the subsequent discovery of the second flavor of neutrino, the muon neutrino (ν_μ). The discovery of ν_μ was achieved in 1962 by a particle accelerator experiment in which highly-energetic muon neutrinos released by decaying π-mesons were guided into a detector. Muon neutrinos have been observed to generate muons but never electrons. Leon Lederman (1922–2018), Melvin Schwartz (1932–2006), and Jack Steinberger (1921–) were awarded the 1988 Nobel Prize in Physics for the discovery of the ν_μ. In the 1970s, particle physicists identified the τ particle. The *tau-neutrino* (ν_τ) and *tau-antineutrino* were discovered in 2000 by physicists at the Fermi National Accelerator Laboratory.

7.7.2 Solar Neutrinos

The Sun is the source of most of the neutrinos that travel around and through us. Approximately 100 billion neutrinos from the Sun go through our thumbnails each second. Neutrinos are produced in the Sun by nuclear fusion: to turn four protons into a helium-4 nucleus requires two protons to be transformed into neutrons by the weak nuclear force, which occurs as two protons fuse into deuterium:

$$p + p \rightarrow d + e^+ + \nu_e. \tag{7.12}$$

This is the principal contribution to the solar neutrino flux. This source of neutrinos is far more prolific than *atmospheric neutrinos* and *diffuse supernova neutrinos*. The maximum solar neutrino flux has a low-energy range of ~400 keV (Figure 7.13). There are also some other essential processes of production, having energy up to ~18 MeV (Bellerive 2004). The primary mechanisms of production of solar neutrinos and their relative fluxes are shown in Figure 7.14.

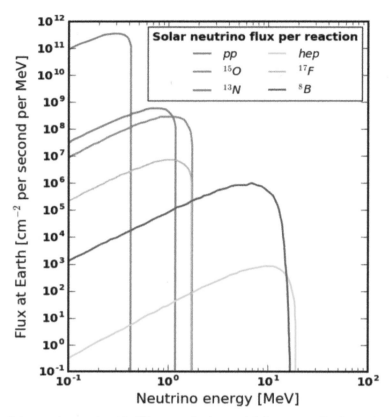

Figure 7.13. Solar neutrinos produced in different mechanisms and their corresponding flux and energy. Image credit: Wikipedia (CC BY-SA 4.0) https://en.wikipedia.org/wiki/Solar_neutrinos.

The Solar Neutrino Problem

When physicists began searching in the 1960s for neutrinos from Sun, a curious thing occurred. The number of neutrinos produced in the core of Sun should be directly related to the number of nuclear reactions (p-p chain and CNO chain). However, only around one-third of the expected number of neutrinos were detected in observations. This is known as the "*Solar neutrino problem*", and it took almost four decades to resolve. It began with the *Homestake experiment*, headed by Ray Davis Jr (1914–2006). Homestake had been an abandoned gold mine and because of its two mile depth, could be used for these scientific studies.

In order to test for neutrinos, the chosen technique used 400,000 l of *perchloro-ethylene*. John Bahcall (1934–2005), Davis' research associate, had estimated how many neutrinos would come from the Sun and convert one of the detector's chlorine atoms into an argon atom. But after examining the data, it appeared as if only one-third of the neutrinos reached the detector. As experimenters refined their methods and gathered more data, and models of the Sun were checked and refined, the issue remained. A more precise Japanese experiment in 1989 called *Kamiokande* only deepened the mystery (also see Section 6.1.3). Their *pure water detector* observed

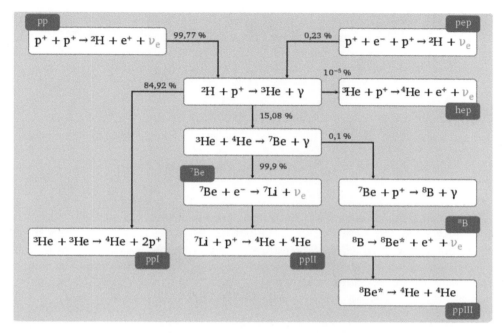

Figure 7.14. Proton–proton chain for solar neutrinos with corresponding branching ratios, showing the various sources of solar neutrinos. Image credit: Wikipedia: Dorottya Szam (CC BY-SA 2.5).

more neutrinos than Davis' experiment by about half of the expected total. However, the problem of lost neutrinos remained.

Neutrino Oscillations

Evidence from two experiments helped to finally resolve the riddle. An updated version of the Kamiokande experiment, known as *Super-Kamiokande*, began observations in 1996. It was followed by the *Sudbury Neutrino Observatory* (SNO) in Canada in 1999.

According to the Standard Model, neutrinos come in three flavors, of which only electron neutrinos are produced in the core of Sun, and detectors were designed to detect only electron neutrinos. In 1968, Bruno Pontecorvo (1913–1993) proposed that neutrinos could change into other flavors if they have at least some mass. Thus, the missing neutrinos had simply changed into other flavors on their way to Earth, rendering them undetectable.

This phenomenon is known as *neutrino oscillation*. The discovery of neutrino oscillations confirmed that neutrinos have a tiny but finite mass.

7.7.3 Atmospheric Neutrinos

Cosmic rays, mostly protons and alpha particles, have typical energies in the GeV energy range, comparable to the rest-mass energy of a proton. The energy spectrum of these particles reaches to very high energies. When these particles interact with nuclei in the upper regions of the atmosphere, exotic high-mass particles such as pions and *K mesons* are formed. These mesons decay into other particles, including neutrinos; for example, a π^+ decays into a muon (μ^+) and a muon neutrino (ν_μ):

$$\pi^+ \rightarrow \mu^+ + \nu_\mu, \tag{7.13}$$

$$\pi^- \rightarrow \mu^- + \bar{\nu}_\mu. \tag{7.14}$$

Muons generated in these reactions are highly unstable and decay to neutrinos:

$$\mu^+ \rightarrow e^+ + \nu_e + \bar{\nu}_\mu, \tag{7.15}$$

$$\mu^- \rightarrow e^- + \bar{\nu}_e + \nu_\mu. \tag{7.16}$$

In this manner, neutrinos are produced when a cosmic ray particle enters the atmosphere. Smaller contributions come from *kaon decays*. Figure 7.15 shows schematically the production of neutrinos in the atmosphere. These neutrinos are called *atmospheric neutrinos*. The above reactions demonstrate that the flux of electron neutrinos and muon neutrinos (and antineutrinos) are related to each other. Flux ratios of these atmospheric neutrino are given in (Kajita 2012):

$$\psi(\bar{\nu}_\mu + \nu_\mu) = 2\psi(\bar{\nu}_e + \nu_e), \tag{7.17}$$

where ψ represents the neutrino flux. Beyond a few GeV, muons on average do not decay before reaching the surface of Earth: $\psi_{\nu_\mu,\bar{\nu}_\mu} > \psi_{\nu_e,\bar{\nu}_e}$. In this instance, the flux

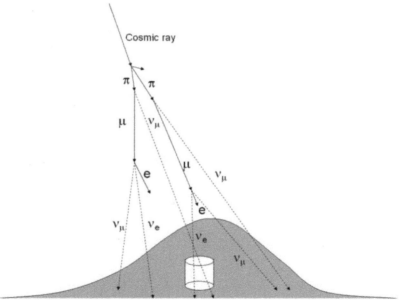

Figure 7.15. A schematic of the production of cascaded atmospheric neutrinos. Image credit: Reprinted from Kajita (2012) with premission from Hindawi.

ratio $\psi_{\nu_\mu,\bar\nu_\mu}/\psi_{\nu_e,\bar\nu_e}$ is still correctly measured because the proportion of muons that decay before reaching the surface can be precisely determined.

7.7.4 Neutrinos from Supernovae

When a star that is 6–8 times more massive than the Sun reaches the end of its supply of nuclear fuel, it explodes in a *supernova*. Much of the mass of the star is ejected, leaving behind a dense "neutron star" remnant with about the mass of the Sun. The extreme energy of the collapse and bounce of the outer layers of the star, including the transformation of the core of the star into pure neutrons, will produce enormous numbers of neutrinos that carry much of the energy of explosion away from the star. The emission of all flavors of neutrinos and antineutrinos creates a high-pressure push on the outer layers of the star, overcoming the gravitational binding energy and creating the expanding ejecta.

This hypothesis was confirmed (Bahcall 1989) by the 1987 discovery of neutrinos released by the SN 1987A supernova, which erupted in the *Large Magellanic Cloud (LMC)*, a small satellite galaxy of our own Galaxy, about 50 kpc away.

7.7.5 Neutrinos from Binary Systems

Binaries are excellent candidates for energetic neutrino generation. As seen in Figure 7.16, a binary consisting of a pulsar and a regular star can generate a strong flux of neutrinos. Around their shared center of mass, the pulsar and the star orbit. If the stellar mass is much larger than the pulsar mass, the pulsar will accelerate protons to very high energies. When these protons interact with the star's atmosphere, pions will be produced, whose subsequent decay produces neutrinos.

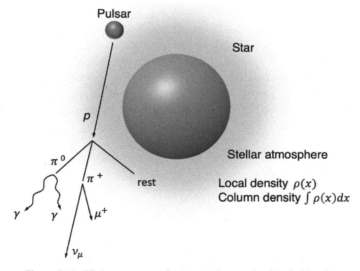

Figure 7.16. High energy neutrino generation mechanism in binaries.

Whether such a source radiates high-energy gamma photons or neutrinos depends on details of the stellar atmosphere, which determines which of the following reactions dominate,

$$\pi^+ \rightarrow \mu^+ + \nu_\mu, \qquad (7.18)$$

$$\pi^- \rightarrow \mu^- + \bar{\nu}_\mu, \qquad (7.19)$$

$$\pi^0 \rightarrow \gamma + \gamma. \qquad (7.20)$$

Further, as the column density of the stellar atmosphere increases, photons will be fully absorbed again. For stellar atmospheres with densities of 10^{-8} g cm^{-3} and column densities beyond 250 g cm^{-2}, this source will only be observable by the emission of neutrinos (Grupen & Buvat 2011).

7.7.6 Detection of Neutrinos

Neutrino Telescopes
Observations indicate that low-energy cosmic rays, which are the most abundant cosmic rays, are produced by the Sun in the *solar wind*. By contrast, the highest energy cosmic rays are believed to be created in high-energy astrophysical sources such as SNRs, GRBs, crashing galaxies, AGNs, and so on. The key to identifying the origin of high-energy cosmic rays is high-energy neutrinos. Since neutrinos are most often the decay products of the muon and kaon, they are a smoking gun of a cosmic proton accelerator, that is, a cosmic ray producer.

IceCube, the world's largest particle detector, was built to detect these high energy neutrinos produced in accelerators billions of light years away. Neutrinos can be indirectly detected after they interact with a target nucleus. The target atom undergoes nuclear transmutation and a charged lepton will appear in place of the neutrino. As these secondary leptons traverse the deep Antarctic ice that makes up the IceCube detector at relativistic speeds, they give off a pale blue light called Čerenkov radiation (see Section 6.1.3). To detect this light, the large, transparent detector must be shielded from daylight. Hence, neutrino observatories are built deep underwater or embedded in ice. Whenever there is a detection of very high energy neutrinos by IceCube, a public alert is sent within a minute to observatories worldwide to look in the direction that the neutrino originated.

Radio-chemical Method
The first approach used to detect low-energy solar neutrinos was the radio-chemical technique. This method relies on inverse beta decay:

$$X^A_N + \nu_e \rightarrow (X + 1)^A_{N-1} + e^-, \qquad (7.21)$$

where $(X + 1)^A_{N-1}$ is highly unstable and decays with a relatively short half-life. Thus, these experiments typically detect the electron neutrino by searching for the decay of a parent nucleus to a daughter nucleus. For example, for a flux of incident neutrinos $\sim 10^{10}$ cm^{-2} s^{-1} and a cross-section of 10^{-45} cm^{-2}, approximately 10^{30} atoms are required to achieve a rate of one occurrence per day. Such experiments are far from simple.

These detectors operate by allowing daughter nuclei to collect over a time that is limited relative to the half-life of the daughter nucleus. The detector is drained regularly and the daughter nuclei collected and measured. This method measures on the rate of interactions of electron neutrinos with the system; it does not measure tau and muon neutrinos, interaction time, neutrino energy, or the direction of the neutrino. The *Gallium Experiment (GALLEX)* and *Gallium Neutrino Observatory (GNO)* used this technique to detect low-energy solar neutrinos.

Čerenkov Water Detectors

Charged particles traveling through a medium at greater than the local speed of light (but not the speed of light in a vacuum, of course) radiate a cone of Čerenkov radiation (see Sections 6.1.3 and 7.4.1). This cone is centered on the particle's direction of propagation and has an opening angle that depends on the particle's velocity and the medium's refractive index:

$$\cos \theta_c = \frac{1}{n\alpha} \tag{7.22}$$

where $\alpha = v/c$ and v is the speed of the particle.

The cone is detected as a ring of light on a sensor. We can reconstruct the trajectory and energy of the particle that created the ring from the impact pattern. Identification of the particles is quite difficult; for example, confusion arises in distinguishing muons from pions. However, from the ring pattern, electron neutrinos can be distinguished from muon neutrinos: instead of the sharp edged circles seen from muon neutrinos, electron neutrinos appear to scatter in the target medium, leading to blurry circles. The two graphs in Figure 7.17 indicate the impact pattern seen with an electron neutrino (top) and muon neutrino (bottom) in the *super-Kamiokande detector*. Since the detection system depends on light passing through a medium, this medium must be extremely transparent. To have an ample target mass, there must also be large amounts of it, so it must be cost-effective. For these types of detectors, the target medium is usually ultra-pure water.

7.8 High-energy Observatories

Here is a list of high-energy observatories or missions, excluding those solely dedicated to observing the Sun.[6]

7.8.1 Active High-energy Astrophysics Missions

- AGILE—A unique combination of a gamma-ray instrument and a hard X-ray imager (2007–present).
- AstroSat—India's first multiwavelength astronomy satellite, with significant X-ray capabilities: 0.3–80 keV (imaging), 10–150 keV (coded mask),

---

[6] Based on https://heasarc.gsfc.nasa.gov/docs/heasarc/missions/active.html and https://heasarc.gsfc.nasa.gov/docs/heasarc/missions/past.html.

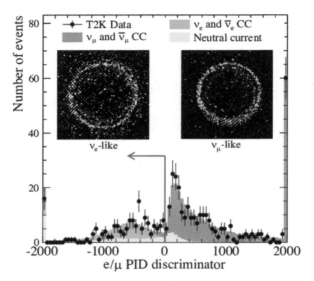

Figure 7.17. Distribution of the parameter used to classify Čerenkov rings in Super-Kamiokande as electron-like (to the left) and muon-like (to the right) together with insets showing examples of an electron-like (left) and muon-like (right) Čerenkov ring. Image from https://commons.wikimedia.org/wiki/File:Superkamiokande_electron_muon_discriminator.png

and 2–10 keV (all-sky monitor; 2015–present). It has ended its five year mission, but continues to operate.

- CALorimetric Electron Telescope (CALET)—An experiment on the International Space Station that investigates possible nearby sources of high-energy electrons, galactic particle acceleration and propagation, and studies gamma-ray bursts (2015–present).
- Chandra—Satellite is in a 64-hour, highly eccentric Earth orbit and has an unprecedented sub-arcsecond X-ray imaging capability as well as both medium-resolution CCDs and high-resolution gratings (1999–present).
- Dark Matter Particle Explorer (DAMPE)—High-energy gamma-ray observatory used to detect electrons and protons (5 GeV–10 TeV), aimed to identify possible Dark Matter signatures, and cosmic ray nuclei up to 100 TeV (2015–present).
- Fermi Gamma-ray Space Telescope—Gamma-ray observatory studies the cosmos in the 30 MeV to 10 GeV range with unprecedented sensitivity and spatial resolution (2008–present).
- HaloSat—CubeSat class mission designed to survey the distribution of hot gas in the Milky Way and constrain the mass and geometry of the Galactic halo. Energy range 0.4–7 keV (2018–present).
- Hard X-ray Modulation Telescope (HXMT), named "Insight," is China's first X-ray astronomy satellite. It aims to scan the Galactic plane to find new transients, monitor variable sources and study gamma-ray bursts. Energy 1–250 keV. (Launched on 2017 June 15, began science operations on 2018 January 30–present.)

- INTEGRAL—The International Gamma-ray Astrophysics Laboratory covers a wide range of X-ray and gamma-ray energies from 3 keV to 10 MeV with high spectral and spatial resolution (2002–present).
- MAXI—A soft X-ray all-sky monitor on board the International Space Station which, when it discovers an X-ray transient, can issue an alert to observers worldwide within 30 s (2009–present).
- NICER—The Neutron star Interior Composition ExploreR is an observatory on the ISS that has large-effective-area concentrator optics in the soft X-ray band, with sub-microsecond timing capability (2017–present).
- NuSTAR—The Nuclear Spectroscopic Telescope Array is the first focusing medium- to high-energy (5–80 keV) X-ray mission with sub-arcminute angular resolution and $\leqslant$ 1.2 keV spectral resolution (2012–present).
- PolarLight is a CubeSat mission to demonstrate a gas pixel detector (GPD) developed for high-sensitivity astronomical X-ray polarimetry (2018–present).
- Swift Gamma-ray Burst mission—The satellite is capable of an autonomous spacecraft response when a GRB is detected, enabling it to obtain accurate position estimates within minutes of the events and to conduct prompt follow-up multiwavelength (X-ray, UV, and optical) observations (2004–present).
- Spectrum–Roentgen–Gamma (SRG)—The primary science of the SRG mission is to carry out the first sensitive all-sky-survey in the energy range 0.3–40 keV. The sky survey is mapped using two instruments: eROSITA covering the lower part on the band in the 0.3–10 keV range and the ART-XC operating in the 4–30 keV (2019–present).
- XMM-Newton—An X-ray observatory in a 48-hour highly eccentric Earth orbit with a very large collecting area and both medium-resolution CCDs and high-resolution gratings as well as the capability to conduct simultaneous X-ray & optical observations (1999–present).

7.8.2 Selected Past High-energy Astrophysics Missions

- *ANS*—Aug 1974–June 1977, Energy range: 0.1–30 keV and 1500–3300 Å.
- Ariel V—Oct 1974–Mar 1980, Energy range: 0.3–40 keV.
- ASCA—1993–2001, Energy range: 0.4–10 keV, First X-ray mission to combine imaging capability with broad pass band, good spectral resolution, and a large effective area.
- BBXRT—Dec 1990, Energy range: 0.3–12 keV, Shuttle-borne instrument.
- BeppoSAX—1996–2002, Energy range: 0.1–300 keV, Broad-band energy coverage and X-ray imaging of the sources associated with gamma-ray bursts and thus determining their positions with an unprecedented precision.
- CGRO (Compton Gamma-ray Observatory)—1991–2000, Energy range: 30 keV–30 GeV, First Great Gamma-ray observatory. Discovery of an isotropic distribution of gamma-ray bursts.
- Copernicus—Aug 1972–late 1980, Energy range: 0.5–10 keV.
- COS-B—Aug 1975–Apr 1982, Energy range: 2 keV–5 GeV.
- DXS—Jan 1993, Energy range: 0.15–0.28 keV, Shuttle-borne instrument.

- Einstein—Nov 1978–Apr 1981, Energy range: 0.2–20 keV.
- EUVE (Extreme Ultraviolet Explorer)—1992–2001, Energy range : 70–760 Å, First dedicated extreme ultraviolet mission.
- EXOSAT—May 1983–Apr 1986, Energy range: 0.05–20 keV, 90 h highly eccentric Earth orbit.
- Ginga—Feb 1987–Nov 1991, Energy range: 1–400 keV.
- Granat—Dec 1989–Nov 1998, Energy range: 2 keV–100 MeV.
- Hitomi—Feb 2016–Mar 2016, Energy range: 0.3–600 keV.
- Hakucho—Feb 1979–Apr 1985, Energy range: 0.1–100 keV.
- HEAO-1—Aug 1977–Jan 1979, Energy range: 0.2–10 keV.
- HEAO-3—Sep 1979–May 1981, Energy range: 50 keV–10 MeV.
- HETE-2—Oct 2000–Oct 2006, Energy range: 0.5–400 keV, designed to detect and localize gamma-ray bursts.
- OSO-7—Sep 1971–Jul 1974, Energy range: 1 keV–10 MeV.
- OSO-8—Jun 1975–Sep 1978, Energy range: 0.15 keV–1 MeV.
- ROSAT (Roentgen Satellite)—1990–1999, Energy range: 0.1–2.5 keV, All-sky survey in the soft X-ray band with catalog containing more than 150,000 objects.
- RXTE—Rossi X-ray Timing Explorer. Dec 1995–Jan 2012, Energy range: 1.5–240 keV, very large collecting area and all-sky soft X-ray monitor, precision timing with 1 microsecond resolution.
- SAS-2—Nov 1972–Jun 1973, Energy range: 20 Mev–1 GeV.
- SAS-3—May 1975–1979, Energy range: 0.1–60 keV.
- Tenma—Feb 1983–late 1984, Energy range: 0.1–60 keV.
- Uhuru—Dec 1970–Mar 1973, Energy range: 2–20 keV.
- Vela 5B—May 1969–Jun 1979, Energy range: 3–750 keV.

7.8.3 Selected Ground Based High-energy Observatories

The Pierre Auger Observatory (PAO) is the world's leading ultra high energy (UHE, $>10^{18}$ eV) cosmic ray observatory. It continuously records data on UHE cosmic rays, which are a mixture of protons and heavier nuclei, arriving over a 3000-km^2 area. The science tasks include measurements of the arrival directions, spectral energy distribution, composition, and interaction properties of these particles in order to understand the extreme environments in which they are accelerated and the intergalactic space through which they propagate.

IceCube is the world's largest neutrino telescope, utilizing a cubic kilometer of ice at the South Pole. Completed in late 2010, the project has had a significant impact in the field of high-energy astrophysics, with the first-ever discovery of high-energy astrophysical neutrinos (at energies up to several PeV) in 2013. It is currently being used to identify neutrino sources and production mechanisms. To date, the coordinates of hundreds of astrophysical neutrinos, reconstructed to tenths-of-degree accuracy, are being used in correlation studies to identify their sources. IceCube also has the ability to detect the strong, low-energy neutrino signals from galactic supernovae, and to detect bursts in the LMC. Further, IceCube sets highly competitive constraints on the parameters of dark matter models via the non-observation of neutrinos from regions where dark matter annihilation is expected, e.g., the Sun, and our own and nearby galaxies. Development is underway for a next-generation detector, *IceCube-Gen2*. Using cosmic ray air shower detectors at the surface, it will improve sensitivity across the full sky by a factor of three for northern hemisphere sources, and increase the number of southern hemisphere detections at 2 PeV by a factor of two. IceCube-Gen2 is anticipated to operate well into the 2040s.

The KM3NeT, is located in the Mediterranean Sea to allow for maximum sensitivity for incoming neutrinos from the southern hemisphere. The neutrinos detected are in the 3 GeV to 10 PeV range.

KM3NeT's ARCA (Astroparticle Research with Cosmics in the Abyss) detector, just off the coast of Sicily, will target high-energy neutrinos in the 100 GeV to 1 PeV range. It is hoped that KM3NeT will identify the source of neutrinos found by IceCube as well as neutrinos suspected from multiple potential particle acceleration sites found by electromagnetic observations. These include Galactic and extragalactic sources such as:

- Supernova remnants,
- Pulsar wind nebulae,
- Sgr A* (a possible PeVatron),
- Fermi bubbles,
- Molecular cloud cosmic ray secondaries,
- AGN,
- Star-forming galaxies,
- GRBs, and
- Cosmic diffuse components.

The primary feature that allows this enhanced source identification is the KM3NeT/ARCA detector's improved angular resolution.

KM3NeT/ORCA (Oscillation Research with Cosmics in the Abyss) off the southern coast of France has a denser configuration designed to study atmospheric neutrino flux in the 3–100 GeV range. Its goals include: the study of neutrino oscillation parameters, the resolution of the neutrino mass hierarchy,[7] probing the nature of dark matter in the GeV to TeV mass range and to provide support nodes for oceanic sciences.

The *CTA* (Figure 1.18) is a international project to build a next-generation ground-based gamma-ray instrument in the energy range extending from some tens of GeV to about 300 TeV. The CTA is the next major step in TeV gamma-ray astronomy. The telescope array will probe extreme particle acceleration (e.g., the lives and death of massive stars, super massive black holes, accretion processes, outflows from galaxy clusters, star formation rates in galaxies, formation of magnetic fields, cosmic rays and electron, influence on the interstellar medium), as well as astroparticle probes of dark matter and even SETI.

---

[7] This is made possible by the sizeable contribution of electron neutrinos to the third neutrino mass eigenstate (a quantum-mechanical state) as reported by An et al. (2012).

The sites for CTA have been selected (Paranal Chile—south; La Palma Canary Islands—north). CTA has entered its pre-production phase, which will realize at least 10 telescopes on the CTA sites (north and south) in early 2022. This will yield the first science and enable further optimization of telescopes, leading into the full production phase. This will see a mass deployment of units, with about 100 telescopes at the southern hemisphere site and about 25 telescopes at the northern site. Approximately 90% of the telescopes are scheduled to be in place by the end of 2025. When completed, CTA will be at least 10 times more sensitive than current TeV gamma-ray facilities such as HAWC, MAGIC, HESS, and VERITAS.

References

Abdalla, H., Abramowski, A., Aharonian, F., et al. 2018, A&A, 612, A1

Abdollahi, S., Acero, F., Ackermann, M., et al. 2020, ApJS, 247, 33

Abeysekara, A. U., Albert, A., Alfaro, R., et al. 2017, ApJ, 843, 40

Acero,, Ackermann, M., Ajello, M., et al. 2015, ApJS, 218, 23

Amelino-Camelia, G., Ellis, J., Mavromatos, N. E., Nanopoulos, D. V., & Sarkar, S. 1998, Natur, 393, 763

An, F. P., An, Q., Bai, J. Z., et al. 2012, NIMPA, 685, 78

Bahcall, J. N. 1989, Neutrino Astrophysics (Cambridge: Cambridge Univ. Press)

Bellerive, A. 2004, IJMPA, 19, 1167

Bergström, L., & Snellman, H. 1988, PhRvD, 37, 3737

Blümer, J., Engel, R., & Hörand el, J. R. 2009, PrPNP, 63, 293

Crocker, R. M., & Aharonian, F. 2011, PhRvL, 106, 101102

De Angelis, A., & Pimenta, M. 2018, Introduction to Particle and Astroparticle Physics: Multimessenger Astronomy and its Particle Physics Foundations (Berlin: Springer)

Degrange, B., & Fontaine, G. 2015, CRPhy, 16, 587

Funk, S., Hinton, J. A., & CTA Consortium, 2013, APh, 43, 348

Grupen, C., & Buvat, I. 2011, Handbook of Particle Detection and Imaging (Berlin: Springer)

H.E.S.S. Collaboration,, Abdalla, H., Abramowski, A., et al. 2018, A&A, 612, A1

Hinton, J. A., & Hofmann, W. 2009, ARA&A, 47, 523

Kajita, T. 2012, AdHEP, 2012, 504715

Klebesadel, R. W., Strong, I. B., & Olson, R. A. 1973, ApJ, 182, L85

Kraushaar, W. L., & Clark, G. W. 1962, PhRvL, 8, 106

Kraushaar, W. L., Clark, G. W., Garmire, G. P., Borken, R., Higbie, P., Leong, V., & Thorsos, T. 1972, ApJ, 177, 341

Morrison, P. 1958, NCim, 7, 858

Nernst, W. 1921, Das Weltgebäude im Lichte der Neueren Forschung (Berlin: Springer)

Petrosian, V., & Bykov, A. M. 2008, SSRv, 134, 207

Selig, M., Vacca, V., Oppermann, N., & Enßlin, T. A. 2015, A&A, 581, A126

Weekes, T. C., Cawley, M. F., Fegan, D. J., et al. 1989, ApJ, 342, 379

Yuan, Q., Yin, P.-F., Wu, X.-F., et al. 2011, ApJ, 730, 15

Chapter 8

Gravitational Waves and Their Production

Gravitational waves allow us to "hear" ripples in the fabric of spacetime itself. In this chapter, we first describe the theory of general relativity, with an emphasis on the description of gravitational waves. We then describe the extreme astrophysical processes that produce gravitational waves, and the methods for their detection.

8.1 What is a Gravitational Wave?

Gravitational waves are propagating fluctuations of gravitational fields, "ripples in spacetime." They can be generated by massive accelerated objects in the universe such as neutron stars and black holes in close orbit with each other. Gravitational waves were predicted by Albert Einstein in 1916 (Einstein 1918) as a consequence of General Relativity (for detailed historical overview see Section 1.3.8). Just as accelerated electrical charge produces electromagnetic waves (light), accelerating masses generate gravitational waves. As with EM waves, spacetime in a gravitational wave oscillates transverse to the direction of propagation. A crucial clue to the origin of gravitational waves comes as a consequence of *Birkhoff's theorem*:[1] a spherically symmetric mass–energy distribution will not radiate gravitational waves. Thus, in the search for sources of gravitational waves, we must look for asymmetry.

The presence of gravitational waves were indirectly inferred in 1982 by the discovery of the orbital decay of the binary neutron star system PSR 1913+16 (Taylor & Weisberg 1982). As the neutron stars orbit each other, their acceleration produces gravitational waves, causing the orbit to decay.

A century after their prediction, gravitational waves were observed directly (Abbott et al. 2016a) for the first time by Advanced Laser Interferometer Gravitational Wave Observatory in 2015. From a multimessenger perspective, LIGO's sixth detection of gravitational waves in 2017 was particularly ground-breaking. It came from a binary neutron star coalescence called GW170817, unlike

---

[1] George David Birkhoff (1884–1944).

previous detections from black hole mergers. GW170817 was followed by a brief burst of gamma rays that were detected with the "Fermi Gamma-ray Burst Monitor" (Goldstein et al. 2017), which established the beginning of multimessenger gravitational wave astronomy. Successive observations established the position of the source and studied its evolution through the electromagnetic spectrum (Mandel 2018).

Gravitational waves are much too feeble to be produced by any mechanism on Earth and subsequently detected. Detectable gravitational wave signals come from energetic astrophysical events, such as binary systems containing black holes or neutron stars. A spinning neutron star would create a periodic gravitational wave signal if the neutron star had an asymmetry that made it non-axisymmetric. Finally, there may be a stochastic background of gravitational waves generated by the superposition of various incoherent sources (Abbott et al. 2018).

To detect a gravitational wave, we must measure changes in distance that are smaller than an atomic nucleus! To do this, we use a *Michelson interferometer*, which uses interfering laser beams to measure small changes in distances between test masses. The test masses are reflecting mirrors positioned at a large distance from each other. The masses are isolated from all perturbations except gravitational waves. Figure 8.1 show schematics of the LIGO facility (Abbott et al. 2016b; Miller & Yunes 2019). The gravitational wave strain or *amplitude h* is defined as:

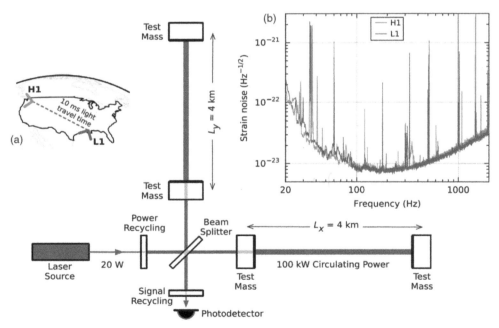

Figure 8.1. A schematic of the Laser Interferometer Gravitational-Wave Observatory (LIGO) facility located in Livingston, Louisiana (United States). Image credit: Reprinted with permissions from Abbott et al. (2016b), Copyright (2016) by the American Physical Society.

$$h \sim \frac{\Delta L}{L}$$

where L and ΔL are the arm length and the small change in arm length in response to gravitational wave stress, respectively. For LIGO, a typical detection will measure $\Delta L \sim 10^{-18}$ m (less than one-thousandth of the diameter of a proton) for a gravitational wave with an amplitude $h \sim 10^{-21}$ and a detector arm length of 4 km. In its quest for gravitational waves, the size of LIGO's instruments is vital. The longer an interferometer's arms, the greater the sensitivity of the instrument, and the smaller the measurements that they can make. Attempting to evaluate a distance shift that is 1000 times smaller than a proton ensures that LIGO must be bigger and much more sensitive than any previously designed interferometer. This is achieved by inserting a mirror to each arm near the beam splitter. Now, the resulting 4 km-long space contains the "Fabry–Perot cavity." After reaching the instrument through the beam splitter, the laser light bounces back and forth 300 times between the mirrors until it is captured in the photodiode (Abbott et al. 2016b). It extends the distance covered by each laser beam from 4 km to 1200 km, thereby raising its arm length effectively. This configuration greatly improves LIGO's sensitivity (Abbott et al. 2016b).

As can be seen in the schematic in Figure 8.1, the operation of a laser interferometer involves sending a laser light beam toward a "beam splitter," which will split the light evenly into two beams that travel along each arm. In normal circumstances (in the case of no gravitational wave signal), the two beams of laser light will reflect off the end mirrors traversing the same distance, return unchanged having no path difference and cancel each other out (Miller & Yunes 2019; Abbott et al. 2016b). Hence, no signal would be detected as no light reached the photo-detector. However, a gravitational wave affects the interferometer differently: one of the arms will extend while the other contracts, and then vice versa as the wave peaks and troughs (also known as half-cycles of the wave; Abbott et al. 2016b). This causes the light to recombine at the output photodetector at slightly different times and create an interference pattern (Miller & Yunes 2019); and hence we can make measurements of the gravitational wave strain or amplitude.

In the following sections, we will discuss the physics of gravitational waves (Section 8.2), the sources and types of gravitational waves (Section 8.2.3), with particular emphasis on the type of gravitational wave known as a "continuous gravitational wave" (Section 8.2.3.4), and gravitational lensing (Section 8.3). This is an exciting new field of astronomy that has captured the public's interest and promises to contribute substantially to our understanding of the universe.

8.2 Physics of Gravitational Waves

8.2.1 Introduction to Gravitational Waves

According to the General Theory of Relativity, gravitational waves will propagate (or ripple) outward from a massive accelerating source such as merging black holes or neutron stars, or perhaps due to the death of a massive star as it undergoes a supernova, and propagate in all directions (Caprini & Figueroa 2018; Figueroa & Torrentí 2017). The wave will travel at the speed of light, i.e., 299,792.458 km s^{-1} (Caprini & Figueroa 2018). In contrast to the impediments encountered by electromagnetic radiation, gravitational waves travel unimpeded due to the fact they are virtually transparent to intervening matter in the universe. They will not be absorbed, scattered, reflected, or altered (Cervantes-Cota et al. 2016; Kokkotas 2002). Thus, the physical information about the source is retained throughout its journey, providing crucial insight into the extreme universe.

Gravitational waves are propagating fluctuations of spacetime (Kokkotas 2002); in which the force generated by the wave will act on the intervening matter perpendicular to the direction of propagation. This interaction has been described by Peter Bergmann (1915–2002, a former collaborator of Albert Einstein; Bergmann 1992) and is shown in Figure 8.2. When a gravitational wave passes through a ring of resting particles, they will successively compress and expand perpendicular to the direction of motion (Rothman 2018; Cervantes-Cota et al. 2016).

As mentioned above in Section 8.1; the source of a strong and detectable gravitational wave is most likely due to the merger or acceleration of extremely massive astronomical objects, such as when a massive binary black hole system *inspirals* and merges. Figure 8.3 shows how an inspiraling binary black hole merger generates gravitational waves. As the two black holes spiral toward each other, they merge and coalesce into one black hole (Abbott et al. 2016b). During this process, the frequency of the gravitational waves being produced will increase significantly until the point of coalescence (Abbott et al. 2016b). After this, the combined black hole undergoes a

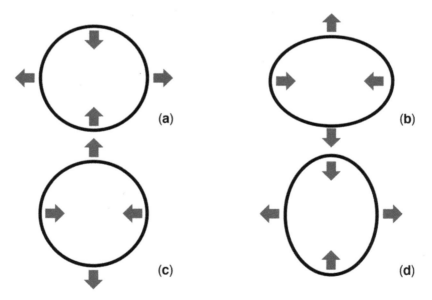

Figure 8.2. A ring of particles at rest will reposition itself according to the directions in a, b, c, d, as a gravitational wave travels orthogonally (transversely) through the particle's space. Or alternately, it can be said that the ring of particles will move transversely to the direction that the gravitational wave travels.

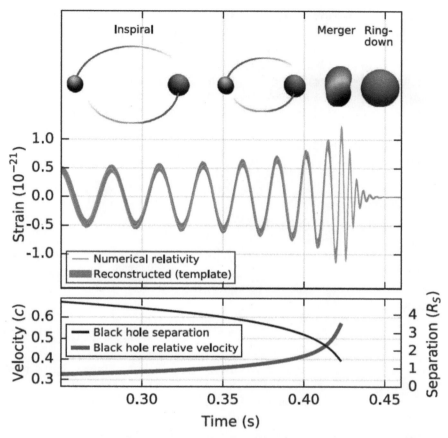

Figure 8.3. As two black holes spiral toward each other and hence reduce their orbital distance, they will accelerate until they coalesce into one black hole, generating strong gravitational waves throughout the final stages (Abbott et al. 2016b). Image credit: Reprinted with permission from Kokkotas (2002), Copyright (2002) by the American Physical Society.

"ring-down" process as it go through a damped oscillation while it settles into an ellipsoidal shape (Abbott et al. 2016b; Kokkotas 2002).

For two bodies orbiting each other in a quasi-circular binary system, the frequency of emitted gravitational waves will be twice the orbital frequency (Amrani 2015). The cycle involves two maxima and two minima (Bondi 1957), hence they are also called "quadrupolar waves."[2] The initial stages of a two-body system depend on their formation mechanism and orbital eccentricity is expected to decay (Kokkotas 2002; Michaely & Perets 2020) and become circular well before they reach the point of coalescence, due to the gravitational radiation produced by an undulating *quadrupole* moment. Interestingly, GW190521 is recently confirmed to be an eccentric merger (Romero-Shaw et al. 2020).

---

[2] It is quadrupolar because one can expand the gravitational waves in spherical harmonics, and the lowest-order mode, which is that we are describing here, is the quadrupolar mode.

Once gravitational waves are generated and begin to propagate throughout space, for the most part, they will simply decrease in amplitude as they spread out but are otherwise unaffected. They experience significant scattering, diffraction, or absorption only if they interact with strong gravitational fields. Such interactions, however rare, can have very interesting effects, such as the formation of *geons*: the waves bunch up and overlap so much that they create their own self-gravitational attraction (Flores et al. 2019). Gravitational waves can also potentially collide with each other and lead to the formation of singularities in the fabric of spacetime (Kokkotas 2002). It remains to be seen whether any such phenomena actually exist in our universe.

8.2.2 The Theory of Gravity

This section will introduce the theory of gravity, first by Newton and then by Einstein, and from this theory derive the predicted properties of gravitational waves.

Gravity was first mathematically described by Isaac Newton in his *Philosophiae Naturalis Principia Mathematica* (1687). He described gravity as a force between any two masses m_A and m_B, which depends on the distance between them r and the Gravitational constant G,

$$F = -\frac{Gm_Am_B}{r^2}. \tag{8.1}$$

The force is always attractive, and acts along the line that connects the two masses. Newton's theory of Gravitation is enormously successful in explaining the effects of gravity on Earth and the movement of bodies in our solar system. Newton did not offer a deeper reason for his equation, famously commenting "I frame no hypotheses" (*Hypotheses non fingo*). Because the effects of gravity travel instantaneously in Newton's theory, there are no Newtonian gravitational waves.

Albert Einstein published his theory of gravity, known as the General Theory of Relativity (GR), in 1915. By describing gravity as the effect of matter and energy on spacetime curvature, Einstein was able to explain and predict phenomena that Newton's theory could not. Within GR, ripples in the curvature of spacetime propagate; these are gravitational waves.

We begin with a brief introduction to GR. In his Special Theory of Relativity (as formulated by Minkowski), Einstein introduces the *spacetime interval ds* between two neighboring points in spacetime, that is, between the infinitesimally-separated events (t, x, y, z) and $(t + dt, x + dx, y + dy, z + dz)$,

$$ds^2 = -c^2dt^2 + dx^2 + dy^2 + dz^2, \tag{8.2}$$

where c is the speed of light. The interpretation is as follows:

- if ds^2 is negative, then $\sqrt{-ds^2}$ is the proper time that ticks on a physical clock that travels inertially between the two infinitesimally-separated events, that is, for whom the events happen at the same location (x, y, z). The interval is known as *time-like*.

- if ds^2 is negative, then light can travel between the two events. The interval is known as *null*. The set of all points that are connected by a null interval to a given event is called the light cone.
- if ds^2 is positive, then $\sqrt{ds^2}$ is the proper distance that will be measured by an inertial observer for whom the two events occurred at the same time. The interval is known as *space-like*.

In GR, the curvature of spacetime is encoded into the relationship between two infinitesimally-separated events and the spacetime interval:

$$ds^2 = g_{\mu\nu} \, dx^\mu \, dx^\nu, \tag{8.3}$$

where $x^\mu = (t, x, y, z)$ for the index $\mu = (0, 1, 2, 3)$, and if an index is repeated on the top and bottom, then it is implicitly summed over. The quantity $g_{\mu\nu}$ is known as the metric tensor.

The metric tensor encodes the curvature of spacetime. The key equation of GR relates the metric tensor (and its derivatives) to the energy contents of spacetime. To understand this equation, we begin by considering what happens to a vector in spacetime as we move around. Imagine holding a stick (a little vector) as you walk around the surface of the Earth, keeping the stick parallel to the ground. Because of the curvature of the Earth, the stick can change its direction, even though we are locally holding it steady. With the benefit of some clever differential geometry, the change in a vector V^α as it travels around a curved surface (known as "parallel transport") is found by solving,

$$\frac{\partial V^\alpha}{\partial x^\beta} + V^\mu \Gamma^\alpha_{\mu\beta} = 0. \tag{8.4}$$

This is called the covariant derivative of V and $\Gamma^\alpha_{\mu\beta}$ is called the *Christoffel symbol*. In a flat space, the *Christoffel symbol* is zero. More generally, the Christoffel symbol is related to the metric tensor as:

$$\Gamma^\gamma_{\beta\mu} = \frac{1}{2} g^{\alpha\gamma} (g_{\alpha\beta,\mu} + g_{\alpha\mu,\beta} - g_{\beta\mu,\alpha}), \tag{8.5}$$

where a comma indicates a partial derivative with respect to the corresponding coordinate index. We can characterize the curvature of a surface by having the vector travel in a closed loop back to where it started, and compare the final vector to the initial vector. We find that if we write down $V^\alpha_{\text{final}} - V^\alpha_{\text{initial}}$ we get:

$$(\Gamma^\alpha_{\beta\nu,\,\mu} - \Gamma^\alpha_{\beta\mu,\,\nu} + \Gamma^\alpha_{\sigma\mu}\Gamma^\sigma_{\beta\nu} - \Gamma^\alpha_{\sigma\nu}\Gamma^\sigma_{\beta\mu}) V^\beta, \tag{8.6}$$

$$= R^\alpha_{\beta\mu\nu} V^\beta, \tag{8.7}$$

where we have defined the *Riemann tensor*, $R^\alpha_{\beta\mu\nu}$, with 20 independent components in a four-dimensional spacetime.

We can create contractions of the Riemann tensor (remembering the summation convention) to define the *Ricci Tensor*,

$$R_{\alpha\beta} = R^{\gamma}_{\alpha\gamma\beta}, \tag{8.8}$$

and the *Ricci Scalar*

$$R = g^{\mu\nu} R_{\mu\nu}. \tag{8.9}$$

We now have the ingredients we need to write down the central *Field equation* of GR, also known as the *Einstein equation*,

$$R_{\alpha\beta} - \frac{1}{2} R g_{\alpha\beta} = \frac{8\pi G}{c^4} T_{\alpha\beta} \tag{8.10}$$

where $T_{\alpha\beta}$ is the *stress–energy tensor*, which describes the distribution of mass–energy. This equation can be motivated in a number of ways; any good GR textbook will take you through at least one approach.

Simplistically, for a given initial distribution of matter and energy in space and time, we solve the Einstein equation to derive the evolution of the spacetime curvature. And since space tells matter how to move, in the words of John Wheeler, this in turn tells us how the matter and energy in the universe evolve (also see Misner et al. 1973).

However, this is a coupled, non-linear set of 10 partial differential equations. Solving the equation is easier said than done! In light of this, we begin by considering simplified cases. If we consider a spherically symmetric, static vacuum scenario, we arrive at the Schwarzschild solution, which describes a black hole. If we consider a homogenous, isotropic scenario, we arrive at the Robertson–Walker metric, which describes an expanding space.

Another obvious simplified case is to consider a small perturbation in flat space (Flanagan & Hughes 2005),

$$g_{\alpha\beta} = \eta_{\alpha\beta} + h_{\alpha\beta}, \tag{8.11}$$

where $\eta_{\alpha\beta}$ is the metric tensor of flat, Minkowski spacetime, $h_{\alpha\beta}$ is a small perturbation such that $\|h_{\alpha\beta}\| \ll 1$ and unless otherwise stated $G = c = 1$ in the derivations below. The smallness of the perturbation means that only linear terms are retained. We can derive the Christoffel coefficients, and from these construct the Riemann tensor, Ricci tensor, and Ricci scalar,

$$R^{\alpha}_{\beta\mu\nu} = \partial_{\mu}\Gamma^{\alpha}_{\beta\nu} - \partial_{\nu}\Gamma^{\alpha}_{\beta\mu} = \frac{1}{2}(\partial_{\mu}\partial_{\beta}h^{\alpha}_{\nu} + \partial_{\nu}\partial^{\alpha}h_{\beta\mu} - \partial_{\mu}\partial^{\alpha}h_{\beta\nu} - \partial_{\nu}\partial_{\beta}h^{\alpha}_{\mu}) \tag{8.12}$$

$$R_{\alpha\beta} = R^{\mu}_{\alpha\mu\beta} = \frac{1}{2}(\partial_{\mu}\partial_{\beta}h^{\mu}_{\alpha} + \partial^{\mu}\partial_{\alpha}h_{\beta\mu} - \Box h_{\alpha\beta} - \partial_{\alpha}\partial_{\beta}h) \tag{8.13}$$

$$R = R^{\alpha}_{\alpha} = (\partial_{\mu}\partial^{\alpha}h^{\mu}_{\alpha} - \Box h), \tag{8.14}$$

where $\Box = \partial_\mu \partial^\mu = \nabla^2 - \partial_t^2$ is the wave or d'Alembertian operator. We can now build the Einstein tensor,

$$G_{\alpha\beta} = \frac{1}{2}(\partial_\mu \partial_\beta h_\alpha^\mu + \partial^\mu \partial_\alpha h_{\beta\mu} - \Box h_{\alpha\beta} - \partial_\alpha \partial_\beta h - \eta_{\alpha\beta} \partial_\mu \partial^\nu h_\nu^\mu + \eta_{\alpha\beta} \Box h). \quad (8.15)$$

If we use the trace-reversed perturbation $\bar{h}_{\alpha\beta} = h_{\alpha\beta} - \frac{1}{2}\eta_{\alpha\beta}h$, all terms with trace h are canceled. We are free to transform our coordinates to our liking; a coordinate transformation known as the Lorentz condition, $\partial^\alpha \bar{h}_{\alpha\beta} = 0$, simplifies the equation to finally be (recalling that $T_{\alpha\beta} = 0$ in a vacuum),

$$G_{\alpha\beta} = -\frac{1}{2}\Box \bar{h}_{\alpha\beta} = 0. \quad (8.16)$$

This is a wave equation, implying that small perturbations in the curvature of empty space ripple outwards.

When applied to the Einstein equation we obtain:

$$\Box \bar{h}_{\alpha\beta} = -\frac{16\pi G T_{\alpha\beta}}{c^4}, \quad (8.17)$$

where $\Box$ is the d'Alembertian Operator and since in a vacuum $T_{\alpha\beta} = 0$.

$$\Box \bar{h}_{\alpha\beta} = 0. \quad (8.18)$$

8.2.3 Sources and Types of Gravitational Waves

Every massive, accelerated, non-spherically symmetric object in the universe produces the fluctuations in the curvature of spacetime called gravitational waves. This includes humans, cars, planes, etc, but the masses and accelerations of these relatively small objects are insufficient to generate detectable gravitational waves. We have to look far outside our own solar system to find detectable signals of gravitational waves.

There are several candidate astronomical events and systems that can produce gravitational waves. Binary system mergers have presented us with the most detections thus far (particularly binary black hole mergers; Abbott et al. 2016b, 2017, 2020). More generally, we can put sources of gravitational waves into (at least) four categories: **compact binary coalescences**, **stochastic**, **burst** and **continuous** (or periodic) gravitational waves.

Compact Binary Coalescence Gravitational Waves
Compact binary coalescence gravitational waves are generated by an orbiting pair of dense and massive objects such as white dwarf stars, neutron stars, and black holes (see Figure 8.4). There are three different classes of "binary compact system"; binary neutron star (BNS), binary black hole (BBH) and neutron star–black hole binary (NS–BH). The gravitational wave signatures from these systems are very difficult to tell apart from one another. However, each binary pair produces a slightly different

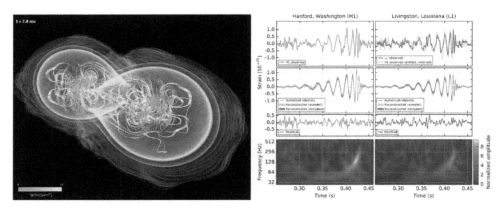

Figure 8.4. (Left) Binary neutron star inspiral. Image credit: Albert Einstein Institute (AEI) & Luciano Rezzolla. (Right) The GW150914 signal observed by LIGO Hanford and Livingston detectors. Image credit: Reprinted with permission from Abbott et al. (2016b), Copyright (2016) by the American Physical Society.

pattern of gravitational waves, but the wave-generation process is the same for all three.

LIGO's instruments track a gravitational wave signal and identify its constituent frequencies, much as human ears are sensitive to a certain set of sound frequencies. LIGO cannot detect objects that are beyond this frequency range. But as the orbiting objects travel closer together, they orbit faster and faster. For much of the merger, binary compact systems can be modeled using Newtonian approximations, considering each neutron star as a point mass. The orbital energy of the binary system is (Hughes 2009),

$$E_{orb} = \frac{G\mu M}{2r},$$

(8.19)

where E_{orb} is orbital energy of the system, $M = m_1 + m_2$ is total mass of system, $\mu = \frac{m_1 m_2}{M}$ is reduced mass, G is gravitational constant, and r is separation between the masses.

The luminosity L of the coalescing system is (Hughes 2009),

$$L = \frac{32G\mu^2 r^4 \Omega^6}{5c^5} = -\frac{dE_{orb}}{dt},$$

(8.20)

where $\Omega = \sqrt{\frac{GM}{r^3}}$ is known as the *orbital frequency*. This gives the expected time until the merger,

$$t = \frac{5c^5 r^4}{256G^3 m_1 m_2 (m_1 + m_2)}.$$

(8.21)

Black hole binaries generate gravitational waves during the coalescences, merger, and ring-down phases. The highest emission amplitude occurs during the merger phase, whose modeling requires numerical relativity techniques (Baker et al. 2006).

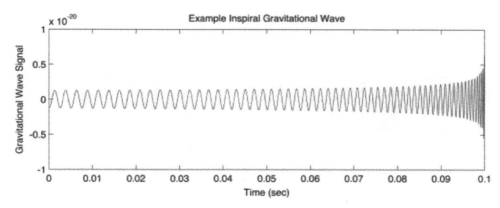

Figure 8.5. This diagram illustrates an example of an inspiral gravitational wave signal. The frequency increases as the bodies within a binary system orbit closer together and eventually coalesce (Abbott et al. 2016b). Image credit: A. Stuver/LIGO Lab/Caltech/MIT.

The first direct observation of gravitational waves, from a binary black hole merger, took place on 2015 September 14 and was announced in cooperation between LIGO and Virgo on 2016 February 11 (Abbott et al. 2016b). The signal was named GW150914, and the source lies at a luminosity distance of 410^{+160}_{-180} Mpc (redshift $z = 0.09^{+0.03}_{-0.04}$). The initial black holes masses were $29^{+5}_{-4}M_\odot$ and $36^{+4}_{-4}M_\odot$ respectively and the final black hole mass is $62^{+4}_{-4}M_\odot$, with $3.0^{+0.5}_{-0.5}M_\odot c^2$ of mass–energy radiated as gravitational waves (Abbott et al. 2016b). Figure 8.5 shows an example of how the wave evolves for a coalescence scenario. The frequency and amplitude increase as the objects approach and merge over millions of years, finally producing an upward-rising "chirp." Beyond the chirp, the wave decays (ringsdown) and the objects settle into a single black hole (Abbott et al. 2016b).

Stochastic Gravitational Waves
Stochastic gravitational waves are theorized to be generated by the superposition of many small, random events in the early universe (Figueroa & Torrentí 2017; Guzzetti et al. 2016).[3] Like the cosmic microwave background, which is thought to be the residual light from the Big Bang (Penzias & Wilson 1965), many sources converge to create a cosmic gravitational wave background. Figure 8.6 shows an example of a stochastic gravitational wave signal.

If detected, this hum could contain crucial details about the origin and earliest moments of the universe. In many inflationary models, they are generated 10^{-36} to 10^{-32} s after the Big Bang (Figueroa & Torrentí 2017); and have since been stretched as the universe expanded. This background hum will be very difficult to detect (Kokkotas 2002).

Detecting gravitational waves from inflation would help to illuminate the universe in the earliest stages (Easther & Lim 2006). For example, Huber & Konstandin (2008) argue that "colliding bubbles in a first-order phase transition"

---

[3] But, they also could come from BBHs, etc.

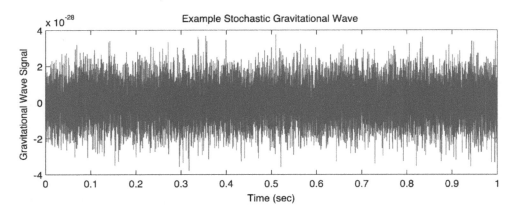

Figure 8.6. An example of a stochastic gravitational wave signal. Image credit: A. Stuver/LIGO Lab/Caltech/MIT.

are potentially responsible for the production of stochastic gravitational waves during the primordial universe; particularly when the first-order phase transitions were ending. They posit that the Higgs[4] field (an energy field that is expected to exist universally) stores kinetic energy that will be gradually released during motions and collisions that occur within the primordial plasma; thereby producing the stochastic waves (Huber & Konstandin 2008). This argument is feasible when one considers how the symmetry of a Higgs field bubble will be disrupted as energy is released during ongoing collisions (Huber & Konstandin 2008); noting again that spherical symmetry cannot generate gravitational waves. More exotic early universe models, such as string cosmology, may also predict a gravitational wave background.

While a stochastic background will be difficult to detect in a single detector, correlation of data from multiple detectors may provide a signal. Through the correlation of the output of detectors displaced thousands of kilometers from one another, LIGO and Virgo can attempt to measure the stochastic background. In the future (perhaps the 2030s), the space-based gravitational wave detector LISA (Ashtekar et al. 2014) will look for a stochastic background from 0.1 mHz to 100 mHz. Pulsar Timing Arrays (in Australia) are searching for this now, albeit in the nHz regime. There are early signs that it might have been seen.

Gravitational Waves from Bursts

Burst gravitational waves comes from short-lived, surprise sources (e.g., supernovae, gravitational waves from magnetar flares, from cosmic strings, etc). Just as new optical, radio or X-ray telescopes can discover astrophysical objects that have never been seen before, we expect gravitational wave detectors to hear as-yet undiscovered phenomena in the universe.

For example, often we don't know enough about the physics of the system to predict how gravitational waves from that source will appear. We're not sure what gravitational waves from supernovae might be like. To hunt in the dark, more

---

[4] After British theoretical physicist Peter Ware Higgs (1929–).

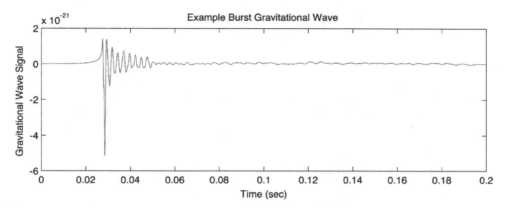

Figure 8.7. This diagram illustrates an example of a burst gravitational wave signal. The frequency and amplitude increase significantly momentarily and then reduce to a stable state. Image credit: A. Stuver/LIGO Lab/Caltech/MIT using data from Ott et al. (2006).

generic algorithms for identifying transient signals have to be used. We cannot limit our analyses to looking only at the signatures of the gravitational waves that we have expected. Additionally (or perhaps alternatively), burst waves could be the result of short-duration phenomena such as gamma-ray bursts (Vladimir & Thorne 1987); an example of a bursty gravitational wave signal is shown in Figure 8.7. The sudden release of energy from a strongly magnetized neutron star (magnetar) is a possible transient source. While "regular" neutron stars are characterized by extremely strong magnetic surface fields ($\sim 10^{12}$ G), magnetars appear to have fields that are 100–1000 times stronger still, suggesting massive magnetic pent-up energy. Soft gamma-ray repeaters (SGRs) and anomalous X-ray pulsars (AXPs) are thought to be separate observational manifestations of the same underlying system—a strongly magnetized star that converts power irregularly from a magnetic field into radiation (Ott et al. 2006).

Gravitational waves may be produced when a massive star ends its life in a supernova explosion. A supernova will only generate gravitational waves if it ejects matter asymmetrically (Kokkotas 2002). This asymmetry could be the result of a supernova in a binary system, or simply the result of an asymmetric explosion.

Hydrodynamic oscillation of the proto-neutron star core is one recently appreciated mechanism for potentially strong gravitational wave emission during core-collapse supernovae (Ott et al. 2006). The energy release implicit for a given source distance and detectable strain amplitude is a general consideration in burst-searching. The energy required to be measurable on Earth for the waves increases as the distance is squared. For a supernova that releases energy E in gravitational waves, with a characteristic duration T and specific frequency f, at a distance r away, we expect a detectable amplitude strain on Earth so long as the energy exceeds (Sathyaprakash & Schutz 2009),

$$E \sim 3 \times 10^{-3} M_{\odot}^2 (h/10^{-21})(T/1 \text{ ms})(f/1 \text{ KHz})(r/10 \text{ Mpc})^2, \qquad (8.22)$$

where h is the strain amplitude of signal, r the supernova distance from Earth that emits energy E in the gravitational wave, timescale T (units are ms) and frequency f.

Continuous Gravitational Waves (CGWs)

The fourth type of gravitational wave source is continuous waves (CWs) produced by compact spinning objects, most notably non-axisymmetric neutron stars in our Galaxy. Despite their relative proximity (several kpc versus tens to hundreds of Mpc for "inspirals"), these sources are expected to have gravitational wave amplitudes that are a few orders of magnitude lower than those seen from compact binary mergers. Our best hope of detecting such weak signals is by aggregating the data over a long period of time. However, this entails significant computational costs due to the broad space of possible signal parameters (e.g., frequency evolution, sky position, possible orbital parameters). Figure 8.8 shows an example of a continuous gravitational wave signal; both the frequency and amplitude are constant. In reality, the frequency and amplitude will vary, but over a much longer timescale than the period of the oscillation.

Figure 8.9 shows an overview of how a continuous gravitational wave signal detected by the LIGO/Virgo facilities will evolve over time; from days to years. Each of the three panels indicates the frequency of the wave (which in this instance is from a neutron star), with the top panel indicating how the frequency of the wave will change in one day as the Earth rotates around its axis; the middle panel indicates how the frequency of the wave will change due to the Earth orbiting the Sun; and the bottom panel indicates how the frequency of the wave will change due to the neutron star losing energy as it rotates, whereby its orbital speed will reduce as time goes on (LIGO/CALweb).

As of 2020, continuous gravitational waves have not been observed. We expect continuous gravitational waves in the frequency band of present ground-based detectors from galactic, non-axisymmetric neutron stars that spin fast enough that their rotational frequencies are observable in the LIGO and Virgo bands. Due to the presence of a time-varying mass quadrupole moment, rotating non-axisymmetric

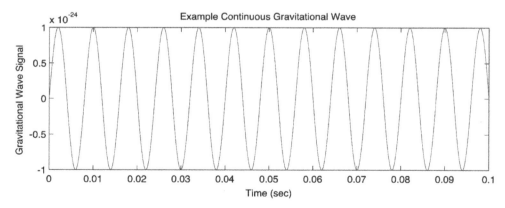

Figure 8.8. Here is an example of a continuous (or periodic) gravitational wave signal. Both the frequency and amplitude remain constant. Image credit: A. Stuver/LIGO Lab/Caltech/MIT.

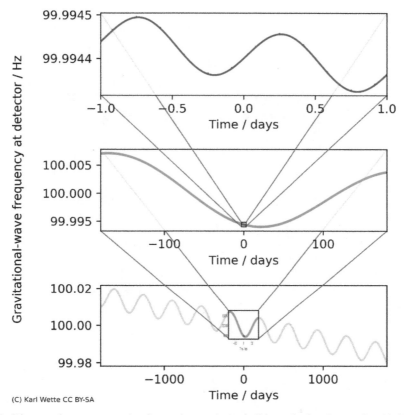

Figure 8.9. Diagram shows an example of a continuous (or periodic) gravitational wave signal being emitted by a neutron star; the top panel indicates the wave signature over one day, the middle panel indicates the wave signature over one year, and the bottom panel indicates the wave signature over many years as the orbital speed of the neutron star decreases (due to the ongoing loss of energy as it rotates). Image credit: Wette, K. (LIGOweb).

neutron stars are expected to emit CGWs. Such signals have smaller characteristic amplitudes than those produced by compact binary mergers, but their overall duration is much longer. In the case of relatively monochromatic CGWs, the integrated signal-to-noise ratio (S/N) increases with the observation time T as:

$$S/N \sim h_o \sqrt{\frac{T}{S}}, \tag{8.23}$$

where the instantaneous GW strain amplitude is h_o and the amplitude spectral density of the frequency of the detector's data signals is S. For example, in the sensitive part of the detector band with the average gravitational wave amplitude $h_o \sim 10^{-21}$, GW150914 lasted for $T \sim 0.2$ s, generating $S/N \sim 24$. The predicted amplitude is a few order of magnitudes smaller for continuous gravitational waves, $h_o \sim 10^{-25}$, but the set of data lasts for T of the order of months or even years (Jaranowski & Krolak 2000).

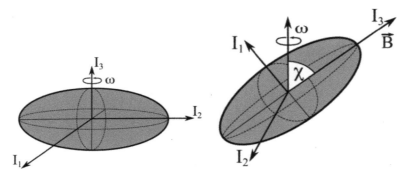

Figure 8.10. (Left) A non-axisymmetric and rotating neutron star (described as a triaxial ellipsoid) radiating gravitational waves at twice the spin frequency. (Right) A model of the CGWs emission due to the magnetic deformation. Image credit: Reproduced from Sieniawska & Bejger (2019) © 2019 by the authors. CC BY 4.0.

Asymmetric deformations in a spinning neutron star can produce continuous gravitational waves. A biaxial body spinning about one of its principle axes (as shown in Figure 8.10) would emit radiation at $f = 2\nu$, where ν is their spin frequency of the neutron star. In this case, gravitational wave strain is (Zimmermann & Szedenits 1979):

$$h_o = \frac{4\pi^2 G I_3 f_{GW}^2}{c^4 d} \tag{8.24}$$

$$= 4.2 \times 10^{-26} \left(\frac{\varepsilon}{10^{-6}}\right)\left(\frac{P}{10 \text{ ms}}\right)^{-2}\left(\frac{1 \text{ kpc}}{d}\right), \tag{8.25}$$

where the ellipticity of the deformation is given by $\varepsilon = \frac{I_1 - I_2}{I_3}$. I_1, I_2, I_3 are moments of inertia in the three major ellipsoid axes, with I_3 associated with the rotation axis of ellipsoid, d is the distance from the source, and P is the period of spin. The total power radiated in the form of gravitational waves from the star is:

$$\frac{dE}{dt} = -\frac{32G}{5c^5} I^2 \varepsilon^2 \Omega^5. \tag{8.26}$$

An absolute upper limit on gravitational wave strain (*spin-down limit*) can be obtained for pulsars by calculating energy loss radiated in the form of gravitational waves. The misalignment between the global dipole magnetic field axis and the rotation axis is responsible for the observed pulsations, interpreted from EM observations. It is possible that an asymmetry in the distribution of the magnetic field within the neutron star may cause the emission of continuous gravitational waves. Chandrasekhar & Fermi (1953) originally proposed the idea that magnetic stress could deform a star and lead to the emission of CGWs (Kalita & Mukhopadhyay 2019).

The model in Figure 8.10(right) was considered in Gal'tsov & Tsvetkov (1984); where the neutron star was modeled by a rigidly spinning Newtonian incompressible fluid body with a uniform internal magnetic field and external magnetic field dipole.

In this model, the inclination of the magnetic dipole moment with respect to the rotation axis is given by the angle χ and the resulting CGWs strain takes the form (Bonazzola & Gourgoulhon 1996),

$$h_o = 6.48 \times 10^{-30} (\beta / \sin^2 \chi)(R/10 \text{ km})^2 (1 \text{ kpc}/d)(1 \text{ ms}/P)(\dot{P}/10^{-13}), \qquad (8.27)$$

where β is the measure of the efficiency of magnetic structure in distorting the star, called the *magnetic distortion factor*. It has been shown that purely poloidal and toroidal magnetic fields are unstable; a realistic model of a neutron star includes mixed field configurations (Braithwaite & Nordlund 2006; Ciolfi et al. 2009; Lander & Jones 2012), which in turn determines the ellipticity:

$$\varepsilon = 4.5 \times 10^{-7} \left(\frac{B_P^2}{10^{15}G} \right) \left(1 - \frac{0.389}{\Lambda} \right), \qquad (8.28)$$

where Λ is the poloidal-to-total magnetic field energy ratio and B_P is the poloidal component of the surface magnetic field of the neutron star. $\Lambda = 0$ for a purely toroidal field and $\Lambda = 1$ for a purely poloidal magnetic field. Neutron stars have magnetic fields less than 10^{15} (Shapiro & Teukolsky 1983). Neutron stars with large magnetic fields of order 10^{15} G are called *magnetars*. Identified magnetars rotate with period (P) of 1–10 s (Manchester et al. 2005); implying that the frequency of the generated continuous gravitational waves will be very low, below the frequency range of Advanced LIGO, Advanced Virgo and other next-generation detectors, such as the Einstein Telescope (Lasky 2015). The right panel in Figure 8.11 shows the estimation of gravitational wave strain for known pulsars observed by the Australia Telescope National Facility (ATNF). The blue crosses represent the purely toroidal field having $\Lambda = 1$, while the green squares represent the strong

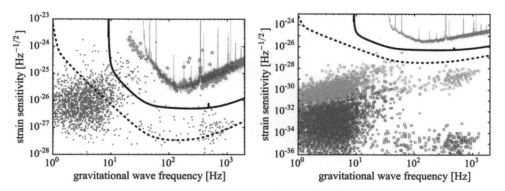

Figure 8.11. (Left) Upper limit on gravitational wave strain from known pulsars (red stars; data from Zimmermann & Szedenits 1979) and spin-down limits for known pulsars in the ATNF catalogue (blue dots). (Right) Gravitational wave strain prediction for known pulsars. Blue crosses and green squares are for normal neutron star matter with purely poloidal magnetic fields and $\Lambda = 0.01$, respectively. Red circles showing the neutron stars are in a color-flavored-locked phase. In both figures, the solid and dashed-black curves indicate the expected sensitivity of the strain for a LIGO and ET, respectively. Image credit: Lasky (2015).

internal toroidal component with $\Lambda = 0.011$. The red circles represent gravitational waves generated by the magnetic field-induced asymmetries in a normal neutron star.

Neutron stars can also generate gravitational waves via internal, seismological oscillations—the analog of earthquakes here. The *Rossby modes* in asteroseismology (known as R-modes, *Carl-Gustaf Arvid Rossby* 1989–1958; Phillips 1998) are internal waves generated by the Coriolis force acting as a restoring force along the surface. R-modes were first proposed as a source of gravitational waves from newborn neutron stars in Owen et al. (1998) and in accreting neutron stars in Bildsten (1998) and Andersson et al. (1999). R-modes have been shown to survive only in a small temperature window where they remain unstable: dissipation at too low temperatures due to shear viscosity dampens the mode, and when the matter is too hot, bulk viscosity will prevent the mode from growing, as shown in the plot in Figure 8.12.

Asymmetric accretion can also lead to asymmetries of composition and heating, which cause stellar deformities (Haensel & Zdunik 1990a, 1990b). Approximate expressions for quadrupolar deformation can be extracted from the quadrupolar portion of temperature variation and reaction threshold energies (Ushomirsky et al. 2000). Such deformations are constrained by the maximum stress that the crust can withstand before breaking (Johnson-McDaniel & Owen 2013), which usually offers higher gravitational wave estimation than the torque balance (Haskell 2015).

Accretion also alters the structure of the magnetic field of the neutron star, which in turn influences the dynamics of accretion. Accreting matter can funnel along magnetic fields along the poles; as the matter settles around the neutron star, it pulls, squeezes, and crushes the local stellar magnetic field (Melatos & Phinney 2001). This mechanism can create larger stellar deformations than those as a result of the background magnetic field inferred from the external dipole (Vigelius & Melatos

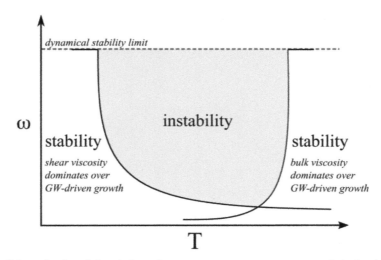

Figure 8.12. Schematic plot of the window of neutron star temperature versus period, showing where the R-modes are unstable and lead to the emission of continuous gravitational waves. The nature of the instability window and the evolutionary NS tracks inside and outside the window are model-dependent. Image credit: Sieniawska & Bejger (2019) © 2019 by the authors. CC BY 4.0.

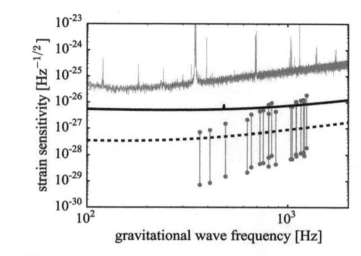

Figure 8.13. Predictions of gravitational waves for magnetic mountains on observed X-ray pulsar accretion with the strength of accretion between fields $B = 10^{10}$ G and 10^{12} G. The solid and dashed-black curves display the predicted sensitivity of the LIGO and ET strains, and the gray curve is the sensitivity of the strain for the initial S5 run. Image credit: Lasky (2015).

2009), and may even produce gravitational waves observable by LIGO or the Einstein Telescope (Priymak et al. 2011), particularly if the buried field is $B > 10^{12}$ (Haskell 2015). This can be seen in Figure 8.13, where the gravitational wave signals from accreting systems with known spin period is seen, assuming initial magnetic fields of $B = 10^{10}$ and 10^{12} G.

8.3 Gravitational Lensing

According to Einstein's theory of gravity (General Relativity), light travels along geodesics (paths of shortest proper distance) in curved, four-dimensional spacetime. Light never stands still, of course, and so gravity bends the path of the beam. This is called *gravitational lensing*. In Figure 8.14, as the photon passes with a distance b (known as the impact parameter) of a compact object with mass M, the local spacetime curvature causes the photon to be deflected by the angle:

$$\alpha = \frac{4GM}{c^2 b} = 1.7'' \left(\frac{M}{M_{\rm Sun}}\right)\left(\frac{b}{R_{\rm Sun}}\right)^{-1}, \tag{8.29}$$

so that, for example, light from a distant star that grazes the surface of the Sun will be deflected by 1.7".

Einstein predicted the deflection of light in 1915. In 1919, the British astrophysicist Arthur Eddington[5] mounted an expedition to photograph stars in the proximity of the Sun during a total solar eclipse (Figure 8.15). Comparison of the eclipse images with images of the same star field taken six months earlier showed that the

---

[5] Sir Arthur Stanley Eddington (1882–1944).

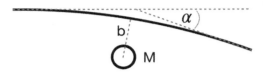

Figure 8.14. Deflection of light by a massive compact object as it passes at impact factor *b*.

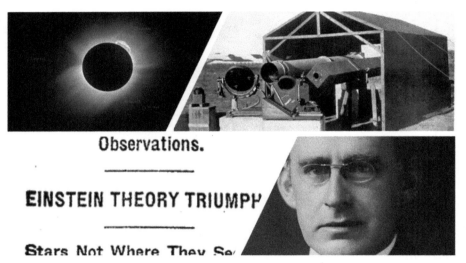

Figure 8.15. Eddington's experiment from 1919. Image credit: (top left) ESO/Landessternwarte Heidelberg-Königstuhl/F. W. Dyson, A. S. Eddington, & C. Davidson (CCBY 4.0); (top right) Wikipedia: C. Davidson; (bottom left) "Einstein theory triumphs"—New York Times, 1919 November 10, p. 17; (bottom right) George Grantham Bain Collection, Library of Congress Prints and Photographs Division Washington, D.C.

positions of the stars had been deflected by the amount that Einstein had expected. This contributed as scientific evidence for the Theory of General Relativity (Lasky 2015).

Galaxy clusters are particularly effective as gravitational lenses. The combined effect of the hundreds to thousands of galaxies in the cluster can create a cosmic magnifying glass enabling us to observe distant galaxies that would otherwise be too faint to see.

Smaller objects, including individual stars, can also serve as gravitational lenses when they pass through the line of sight of more distant stars. For a few days or weeks, the more distant star appears brighter. This effect, where the effect of the lensing is to increase the brightness of the object rather than observably distort the image, is referred to as gravitational *microlensing*.

Gravitational waves will also be deflected by gravitational fields in the same manner as ordinary electromagnetic radiation (Meena & Bagla 2020).[6] Meena &

---

[6] For example, the signals produced by merging binary systems. However, all gravitational waves will travel along geodesics. They will all get bent the same amount, independent of the source or frequency of the gravitational wave.

Bagla (2020) argue that the latter will cause non-zero time delays in the signal (also referred to by the authors as a time-varying phase shift), which can result in detectable signal variations for the range of images that strong gravitational lensing is expected to produce. They posit however that the level of uncertainty generated by microlensing will likely be negligible, albeit should still be considered when parameterizing assessments of each candidate gravitational wave source (Meena & Bagla 2020).

The shapes of distant galaxies can also be distorted by gravitational lensing. Where this distortion is a mere perturbation—a slight elongation on one axis and contraction along another—this effect is known as *weak lensing*. Mass estimates of the clusters can be gained by statistically combining these minor distortions, providing a powerful handle on the large-scale distribution of mass in the universe. Present and future observations will use this effect to study the large scale structure of the universe, from which we can infer its dark matter and dark energy contents.

This shape distortion can also be much more dramatic than a mere perturbation. Like light passing through textured glass, gravitational lensing can distort even compact point sources into arcs and rings of light. This is known as *strong lensing*. The doubly imaged quasar Q0957+561 by Walsh et al. (1979) was the first strong lens to be observed (Figure 8.16, left). This quasar was thought to be two distinct objects when it was discovered in 1979. Detailed observations later revealed that these twins were a little too similar to be a mere coincidence! They are, in fact, the same object, whose light has reached us via two distinct bent paths.

A spectacular example of strong lensing can be seen in the Galaxy ESO 325–G004, which shows light from a background galaxy bent into an "Einstein ring" (Collett et al. 2018). By measuring both the galaxy's mass and spatial curvature of the foreground lens, they were able to confirm general relativity at an extragalactic scale.

How much will gravitational waves be affected by lensing as they travel through the universe? Li et al. (2018) explored strong gravitational lensing of gravitational waves produced via binary stellar-mass black hole mergers, in the context of the expectations of the LIGO observatory at its current level of sensitivity. They predict approximately one lensed gravitational wave event per year. Figure 8.17

Figure 8.16. (Left) Hubble Space Telescope image of QSO 0957+561 showing the effect of strong lensing. Image credit: ESA/Hubble & NASA. (Right) Gravitational lensing in the Abell 2218 galaxy cluster, imaged by HST. Image credit: NASA, Andrew Fruchter and the ERO Team (STScI).

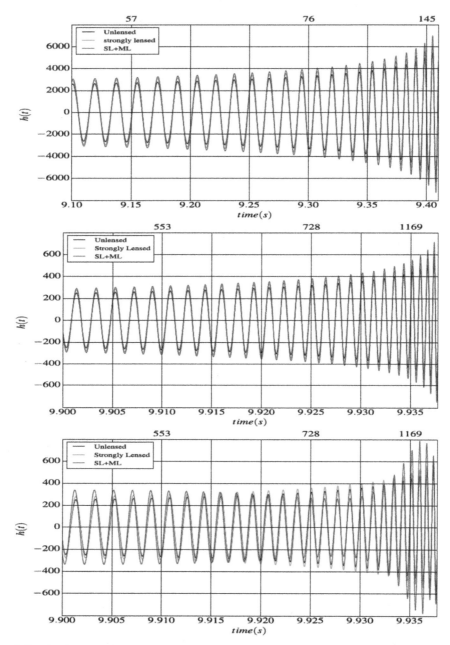

Figure 8.17. The three panels illustrate the effect of strong gravitational lensing (red line), and combined strong lensing and microlensing (green line), on an unlensed gravitational waveform (black line) for three binary systems that are about to coalesce. The top panel shows waveforms for a binary $10\ M_\odot + 10\ M_\odot$ merger; the middle and bottom panels show a $1\ M_\odot + 1\ M_\odot$ merger; the bottom panel also includes a $20\ M_\odot$ microlens. Image credit: Reproduced from Meena & Bagla (2020), Copyright 2020, OUP.

shows the predicted impact of strong gravitational lensing (red line), and combined strong lensing and microlensing (green line), on an unlensed gravitational waveform (black line) for three binary systems that are coalescing. The top panel shows waveforms for a binary 10 $M_\odot$ + 10 $M_\odot$ merger; the middle and bottom panels show a 1 $M_\odot$ + 1 $M_\odot$ merger; the bottom panel also includes a 20 $M_\odot$ microlens (Meena & Bagla 2020).

We see that strong lensing amplifies the waveform. The combined strong lensing and microlensing waveform in the lower panel is distinct; there is a non-uniform amplification along the unlensed waveform.

8.4 Gravitational Wave Observatories

Ever since gravitational waves were first predicted by Einstein in 1916, efforts have been underway to detect them. These efforts have accelerated in the past 50 years, as experimenters/observers have tried to grapple with the extraordinary challenges of detection. Gravitational waves are extremely feeble: gravitational waves passing the Earth produce differential motion on the order 10^{-18} m in a LIGO-size instrument (Figure 8.18); for comparison, a single proton is roughly 10^{-15} m in size. Further, their amplitude decreases as the inverse of source distance.

The three main classes of detector use resonant mass, pulsar timing and laser interferometers to detect gravitational waves. We now discuss these in turn.

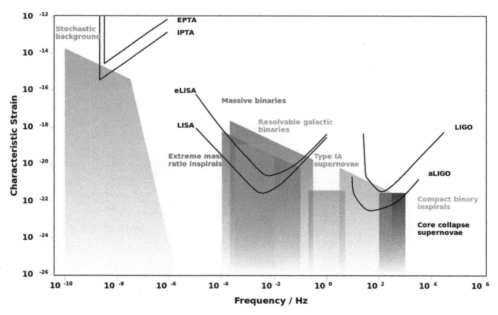

Figure 8.18. Typical noise curves of various gravitational-wave detectors, as a function of frequency and the characteristic strain, for a range of astrophysical sources. Image credit: Wikipedia: Christopher Moore, Robert Cole and Christopher Berry, (CC BY-SA 1.0).

8.4.1 Resonant Mass

A resonant mass antenna is an isolated solid body of metal that attempts to detect gravitational waves via the body's resonant frequency. This was the first method used to look for gravitational waves. It was hoped that a nearby supernova might be strong enough to be detected. Despite a long and convoluted history of claims of detection, none have been verified as of 2021. Given the tiny amplitude of gravitational waves detected by LIGO, it seems that astrophysical events capable of being observed by a resonant mass experiment are very rare.

Active detectors include: NAUTILUS (IGEC), AURIGA (Antenna Ultracriogenica Risonante per l'Indagine Gravitazionale Astronomica[7]) MiniGRAIL, and Mario Schenberg (Brazilian Graviton Project). Past projects include The Weber Bar, EXPLORER, ALLEGRO, NIOBE, ALTAIR, and GEOGRAV.

8.4.2 Pulsar Timing

Pulsars are incredibly accurate clocks. Observations that closely monitor pulsar periods could infer the effect of a passing gravitational wave. This is because the time of arrival of pulsar signals from different directions are shifted rippling of spacetime. This would allow the detection of gravitational waves in the nanohertz range, which is the same frequencies expected to be emitted by pairs of merging supermassive black holes.

The International Pulsar Timing Array (IPTA) formed in 2005 is a multi-institutional, multitelescope collaboration (Hobbs et al. 2010), comprised of the European Pulsar Timing Array (EPTA), the North American Nanohertz Observatory for Gravitational Waves (NANOGrav), and the Parkes Pulsar Timing Array (PPTA). Combining resources and using an array of about 30 pulsars, the goal is to detect and record gravitational wave data that can be shared by each of the participating institutions.

8.4.3 Interferometers

Interferometric gravitational-wave detectors are the only instruments to date that have been successful in detecting gravitational radiation. There can be grouped into generations based on the technology used (Punturo et al. 2010; Gregory 2012). The first generation of detectors were built in the 1990s and early 2000s. They were not expected to detect gravitational waves, but rather to develop the technology necessary to allow detection using more advanced instruments (Punturo et al. 2010; Gregory 2012).

The second generation of detectors used more sophisticated detection methods, including cryogenic mirrors and the injection of a compressed vacuum (Gregory 2012). The result of this effort was the first unambiguous detection of a gravitational

---

[7] http://www.auriga.lnl.infn.it

wave by Advanced LIGO in 2015. A third generation of detectors are now in the planning phase.

The field of gravitational wave astronomy is in its infancy. Detector technology will undoubtedly improve in future decades. Past, current and planned gravitational wave interferometer detectors include:

1. First generation: TAMA 300 (1995), GEO 600 (1995), LIGO (2002), CLIO (2006), and Virgo interferometer (2007),
2. Second generation: GEO High Frequency (2010), Advanced LIGO (2015), Advanced Virgo (2016), KAGRA (LCGT)(2019), LIGO-India (2023), AIGO,
3. Third generation: Einstein Telescope (2030s), Cosmic Explorer (2030s), and
4. Third generation, space-based: TianQin (2035), Taiji (gravitational wave observatory; 2030s?), Deci-hertz Interferometer Gravitational wave Observatory (DECIGO; 2027), Laser Interferometer Space Antenna (Planned for 2030s; The Lisa Pathfinder, a development mission, was launched December 2015) and Neutron Star Extreme Matter Observatory (NEMO; Ackley et al. 2020).

References

Abbott, B. P., Abbott, R., Abbott, T. D., et al. 2016a, PhRvD, 93, 122003

Abbott, B. P., Abbott, R., Abbott, T. D., et al. 2016b, PhRvL, 116, 061102

Abbott, B. P., Abbott, R., Abbott, T. D., et al. 2017, PhRvL, 119, 161101

Abbott, B. P., Abbott, R., Abbott, T. D., et al. 2018, PhRvL, 120, 091101

Abbott, R., Abbott, T. D., Abraham, S., et al. 2020, PhRvL, 125, 101102

Ackley, K., Adya, V. B., Agrawal, P., et al. 2020, PASA, 37, e047

Amrani, D. 2015, PhyEd, 50, 142

Andersson, N., Kokkotas, K. D., & Stergioulas, N. 1999, ApJ, 516, 307

Ashtekar, A., Berger, B. K., Isenberg, J., & MacCallum, M. A. H. 2014, arXiv:1409.5823

Baker, J. G., Centrella, J., Choi, D.-I., Koppitz, M., & van Meter, J. 2006, PhRvL, 96, 111102

Bergmann, P. G. 1992, The Riddle of Gravitation (North Chelmsford, MA: Courier Corporation)

Bildsten, L. 1998, ApJL, 501, L89

Bonazzola, S., & Gourgoulhon, E. 1996, A&A, 312, 675

Bondi, H. 1957, Natur, 179, 1072

Braginsky, V. B., & Thorne, K. S. 1987, Natur, 327, 123

Braithwaite, J., & Nordlund, Å. 2006, A&A, 450, 1077

Caprini, C., & Figueroa, D. G. 2018, CQGra, 35, 163001

Cervantes-Cota, J., Galindo-Uribarri, S., & Smoot, G. 2016, Univ, 2, 22

Chandrasekhar, S., & Fermi, E. 1953, ApJ, 118, 116

Ciolfi, R., Ferrari, V., Gualtieri, L., & Pons, J. A. 2009, MNRAS, 397, 913

Collett, T. E., Oldham, L. J., Smith, R. J., et al. 2018, Sci, 360, 1342

Easther, R., & Lim, E. A. 2006, JCAP, 2006, 010

Einstein, A. 1918, SPAW, 154

Figueroa, D. G., & Torrentí, F. 2017, JCAP, 2017, 057

Flanagan, É. É., & Hughes, S. A. 2005, NJPh, 7, 204

Flores, C. V., Parisi, A., Chen, C.-S., & Lugones, G. 2019, JCAP, 2019, 051

Gal'tsov, D. V., & Tsvetkov, V. P. 1984, PhLA, 103, 193

Goldstein, A., Veres, P., Burns, E., et al. 2017, ApJL, 84, L14

Guzzetti, M. C., Bartolo, N., Liguori, M., & Matarrese, S. 2016, NCimR, 39, 399

Haensel, P., & Zdunik, J. L. 1990a, A&A, 227, 431

Haensel, P., & Zdunik, J. L. 1990b, A&A, 229, 117

Harry, G.M. 2012, in The Twelfth Marcel Grossmann Meeting: On Recent Developments in Theoretical and Experimental General Relativity, Astrophysics and Relativistic Field Theories (in 3 Volumes), ed. A.H. Chamseddine (Singapore: World Scientific), 628

Haskell, B. 2015, IJMPE, 24, 1541007

Hobbs, G., Archibald, A., Arzoumanian, Z., et al. 2010, CQGra, 27, 084013

Huber, S. J., & Konstandin, T. 2008, JCAP, 2008, 022

Hughes, S. A. 2009, ARA&A, 47, 107

Krolak, A., & Jaranowski, P. 2000, PhRvD, 61, 062001

Johnson-McDaniel, N. K., & Owen, B. J. 2013, PhRvD, 88, 044004

Kalita, S., & Mukhopadhyay, B. 2019, MNRAS, 490, 2692

Kokkotas, K. D. 2002, Encyclopedia of Physical Science and Technology, 7, 67

Lander, S. K., & Jones, D. I. 2012, MNRAS, 424, 482

Lasky, P. D. 2015, PASA, 32, e034

Li, S.-S., Mao, S., Zhao, Y., & Lu, Y. 2018, MNRAS, 476, 2220

Manchester, R. N., Hobbs, G. B., Teoh, A., & Hobbs, M. 2005, AJ, 129, 1993

Mandel, I. 2018, ApJL, 853, L12

Meena, A. K., & Bagla, J. S. 2020, MNRAS, 492, 1127

Melatos, A., & Phinney, E. S. 2001, PASA, 18, 421

Michaely, E., & Perets, H. B. 2020, MNRAS, 498, 4924

Miller, M. C., & Yunes, N. 2019, Natur, 568, 469

Misner, C. W., Thorne, K. S., & Wheeler, J. A. 1973, Gravitation

Ott, C. D., Burrows, A., Dessart, L., & Livne, E. 2006, PhRvL, 96, 201102

Owen, B. J., Lindblom, L., Cutler, C., et al. 1998, PhRvD, 58, 084020

Penzias, A. A., & Wilson, R. W. 1965, ApJ, 142, 419

Phillips, N. A. 1998, BAMS, 79, 1097

Priymak, M., Melatos, A., & Payne, D. J. B. 2011, MNRAS, 417, 2696

Punturo, M., Abernathy, M., Acernese, F., et al. 2010, CQGra, 27, 084007

Romero-Shaw, I., Lasky, P. D., Thrane, E., & Bustillo, J. C. 2020, ApJL, 903, L5

Rothman, T. 2018, AmSci, 106, 96

Sathyaprakash, B. S., & Schutz, B. F. 2009, LRR, 12, 2

Shapiro, S. L., & Teukolsky, S. A. 1983, Black Holes, White Dwarfs, and Neutron Stars: The Physics of Compact Objects (New York: Wiley)

Sieniawska, M., & Bejger, M. 2019, Univ, 5, 217

Taylor, J. H., & Weisberg, J. M. 1982, ApJ, 253, 908

Ushomirsky, G., Cutler, C., & Bildsten, L. 2000, MNRAS, 319, 902

Vigelius, M., & Melatos, A. 2009, MNRAS, 395, 1972

Walsh, D., Carswell, R. F., & Weymann, R. J. 1979, Natur, 279, 381

Wang, Y.-B., Zhou, X., Wang, N., & Liu, X.-W. 2019, RAA, 19, 030

Zimmermann, M., & Szedenits, E. Jr 1979, PhRvD, 20, 351

Principles of Multimessenger Astronomy

Miroslav D Filipović and Nicholas F H Tothill

Chapter 9

Obtaining and Interpreting Astronomical Data

Obtaining and defining quantitative data is what makes astronomy a science capable of understanding the universe. In this chapter, we describe coordinate systems and measurement definitions, along with the physics of the detectors necessary to obtain data. Storage and access are also critical to the communities of astronomers that study it.

9.1 Celestial Coordinates

Coordinate systems are a way to define points in space. For any given coordinate system, a set of numbers uniquely defines a point. Moreover, the coordinate system allows a distance between points to be calculated, and, as more dimensions are added, areas and volumes can also be found. The physical universe can be represented by many different coordinate systems, all of which can be converted into each other. The selection of coordinate system is a matter of convenience—very often the symmetry of a particular spatial phenomenon will suggest a coordinate system. In the case of astronomical observations, the most convenient coordinate systems are generally some form of polar coordinates.

9.1.1 General Coordinate Systems

Cartesian Coordinates: The Cartesian coordinate system describes the position of a point in space by a set of numbers measuring the position against a set of axes at right angles to one another. To measure a three-dimensional space, three axes are required, conventionally defined as x-, y-, and z-axes (Figure 9.1; Rogers 2008).

   Polar Coordinates: In situations with radial symmetry about a special point, polar coordinates are often more convenient. In two dimensions, a point is defined by its distance from the central point (conventionally r) and the angle between the line

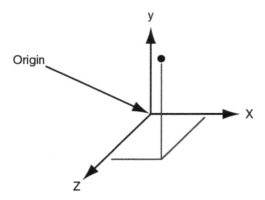

Figure 9.1. This Cartesian coordinate system has three axes: *x*, *y*, and *z*, in addition to the point of origin.

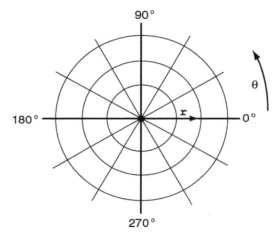

Figure 9.2. A basic overview of a two-dimensional polar coordinate system.

joining that point to the center and a reference direction (Figure 9.2; Rogers 2008).[1] This angle (conventionally θ) is also known as the *azimuth*.

In order to map a three-dimensional space, another coordinate must be added to (r, θ). An axis could be added at right angles to the plane of the original two-dimensional polar coordinates (*z*), with the extra coordinate being just the height above the plane, giving a system of *cylindrical polar coordinates*.

More often, the third coordinate is also angular. The axis at right angles to the original plane (*polar axis*) is retained, but the height above the plane is given instead either by the angle between the line joining the point to the center and the plane or the angle between the line and the pole (Figure 9.3). This angle is generally written θ,

[1] Oregon state university online page, http://sites.science.oregonstate.edu/math/home/programs/undergrad/CalculusQuestStudyGuides/vcalc/coord/coord.html.

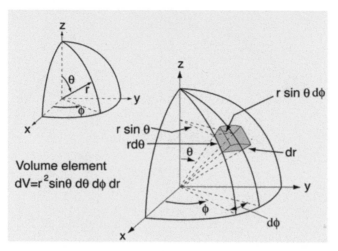

Figure 9.3. An overview of the spherical polar coordinate system, which applies a radial coordinate (r), in addition to the polar angular coordinate (θ). Image credit: HyperPhysics (© C.R. Nave, 2017)[2].

and the angle around the polar axis (θ in two-dimensional polar coordinates) is written ϕ.[2] This gives a system of *spherical polar coordinates*.

9.1.2 Celestial Coordinate Systems

The Celestial Sphere

When we observe the universe from the Earth, we have many lines of sight available to us, all starting at the Earth and going out radially into space. Such a situation naturally suggests the use of spherical polar coordinates. Celestial coordinate systems are a modification of spherical polar coordinates, from which the radial coordinate r has been removed, leaving only the angular coordinates.

While this implies that the coordinate system no longer uniquely identifies a point in space, this reflects the reality of astronomical observation. While the line of sight toward an astrophysical object can be measured directly, its radial distance cannot be, and must instead be inferred from other measurements. So, for the practical purpose of observing the universe, r is unnecessary. Neglecting r has the same effect as assuming that all objects being observed are at the same r, and this gives rise to the idea of the celestial sphere. This is an imaginary sphere, centered on the Earth, of a large size. All of our lines of sight lead out to the inside of this sphere.

Because there is no radial coordinate, the system is actually two-dimensional. The distance between two points is the angle between the radial lines to those points, measured in angular units, such as degrees or radians. Many contiguous lines of sight can define a solid angle (rather than an area), measured in units such as steradians. No volume can be defined in these coordinates.

---

[2] Hyperphysics online page, http://hyperphysics.phy-astr.gsu.edu/hbase/sphc.html#c3.

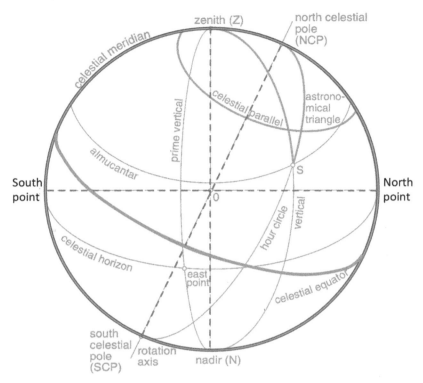

Figure 9.4. This set of labels on the celestial sphere refer to equatorial and horizontal coordinates, which are defined by the Earth's rotation axis, and the plumb line (or vertical reference line), respectively.

Many different systems of spherical polar coordinates can be drawn on the celestial sphere. Since the central origin of the system (i.e., the Earth) does not change, each coordinate system is defined by a plane, around which ϕ is measured. This plane can in turn be defined by its poles, which are the two directions along the axis perpendicular to the plane (see e.g., Figure 9.4).

Angular Measurement: The angular distance between two points on the celestial sphere can be expressed in a variety of units, all of which are based on dividing up a circle. The most familiar way of doing this is in degrees, where the whole of a circle contains 360°. For the finer measurements often used in astronomy, *decimal degrees* may be used (e.g., 34.29843°, with any number of decimal places). *Sexagesimal notation* may also be used, in which one degree is divided into 60′ (or minutes of arc). Each arcminute is in turn divided into 60″ (or seconds of arc), so a degree is equivalent to 3600″. If greater precision is required, decimal arcseconds are generally used.[3] Radians may also be used, where a full circle is made up of 2π radians. It is also possible to state an angle as simply a fraction of a circle; and also as hours, minutes and seconds of time. This latter usage is only for the *right ascension*

---

[3] A sexagesimal angle might be written as $23° \, 36^m 48.2^s$, or $23° \, 36′ \, 48.2″$. Both denote an angle of 23°, 36 arcmin, and 48.2 arcsec.

coordinate in the equatorial system—but it is ubiquitous in astronomy: the circle is divided into 24 h, which are then subdivided into minutes and seconds.

The History of Celestial Coordinates

The celestial coordinate system has a remarkable history that dates back to the times of antiquity; challenging and inspiring philosophers, astronomers and mathematicians throughout its evolution (Figure 9.5 illustrates one such account of inspiration; Flammarion[4] depicted the celestial sphere in his 1888 book, *L'atmosphère: Météorologie Populaire*; Birney et al. 2006). Parmenides of Elea[5] is credited as being the first to discover the Earth's actual sphericity in the early fifth century

Figure 9.5. The famous woodcut depiction of the celestial sphere and "man's quest for knowledge and understanding of the Universe" (Capova et al. 2018), by Nicolas Camille Flammarion in his 1888 book *L'atmosphère: Météorologie Populaire*. Image credit: Camille Flammarion's *L'atmosphère: Météorologie Populaire* in 1888; Birney et al. (2006).

---

[4] Nicolas Camille Flammarion (1842–1925), French astronomer.
[5] Greek philosopher, who lived in southern Italy.

BCE; he also discovered that the Sun is responsible for illuminating our Moon (North 2008). The Greek philosophers Empedocles and Anaxagoras discovered how solar eclipses occur (noting it was the Moon that obscured the Sun), and philosopher and atomic theorist Democritus was developing star catalogs.

In the fourth century BCE, the celestial coordinate system was beginning to emerge due to further progress by Eudoxus of Cnidos.[6] By the end of the fourth century, spherical astronomy had been developed by two of the great ancient astronomers, Autolycus and Euclid (North 2008). Autolycus of Pitane (modern-day Turkey) was responsible for works on the geometry of the celestial sphere, while Euclid of Alexandria was renowned for his geometric work known as *The Elements*.

The Greek astronomer, mathematician and geographer Hipparchus (190–120 BCE) cataloged stars and listed their positions using a "consistent Greek scheme of coordinates" (North 2008). It seems that Hipparchus was using the equatorial system of right ascension and declination; Ptolemy[7] indicates in the highly influential *Almagest* that Hipparchus may have moved to ecliptic coordinates in his work, due to his discovery of the precession of the equinoxes (Dennis 2002).

Around ~500 BCE, the Babylonians developed a numerical system of ecliptic longitudes, a great improvement on earlier reference systems in which stars were assigned to zodiacal constellations (North 2008). Approximately 600 years later, Ptolemy introduced a zero-point for longitudes—the point at which the ecliptic meets the equator (North 2008). Ptolemy's *Almagest* was a masterful exposition of the astronomy known in his time, essentially a study of celestial motions (Woolard 1942); it dominated astronomy for approximately 1400 years, whereupon Nicholas Copernicus (1473–1543); and others, such as Tycho Brahe—who was the first to employ both ecliptic and equatorial coordinate systems to determine a cometary trajectory (North 2008), challenged ancient thought and revolutionized astronomy. While the heliocentric model of solar system was understood and adopted, celestial coordinates remained the same. Since we still observe from the Earth, we observe the universe in a geocentric system.

Equatorial Coordinates

If a spherical coordinate system can be defined by a plane and its associated poles (see above), one of the most natural systems would be to use the poles of the Earth's rotation—the North and South poles—and the plane of its rotation—the Equator. This generates equatorial coordinates, the dominant system used in observational astronomy. The northern and southern celestial poles lie on the axis of the Earth's rotation, and the celestial equator is a projection of the Earth's equator outwards into the universe. The coordinates used are the *right ascension*, a longitude-like coordinate usually written as RA or α, and the *declination*, a latitude-like coordinate usually written as Dec or δ (Figure 9.6).

---

[6] In modern-day Turkey, lived between 400 and 347 BCE and developed a planetary theory that employed geometry to describe spherical motions.
[7] Claudius Ptolemy (100–170 CE), Graeco–Egyptian astronomer.

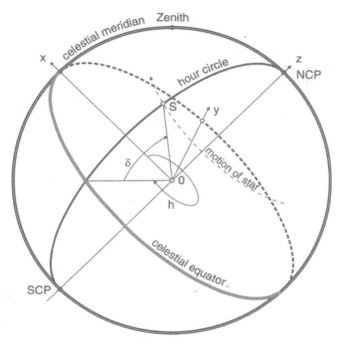

Figure 9.6. An overview of the equatorial coordinate system. If the intersection of the x-axis is the vernal equinox, the angle labeled h is the right ascension.

Right ascension increases to the East, and its zero-point is the "first point of Aries," the position of the Sun in equatorial coordinates at the moment of the vernal Equinox around March 21 each year; this is one of the two points where the path of the Sun in the sky (the *ecliptic*) crosses the celestial equator. Because the zero-point of right ascension is set by a celestial object, not a reference point on the Earth, the equatorial coordinate grid does not rotate with the Earth. Stars in the sky have fixed coordinates, but they, and the coordinate grid, seem to move across the sky with time as the Earth rotates.

Declination is measured from the equator northwards to the north celestial pole (as a positive declination) and southwards to the south celestial pole (as a negative declination).

Definition and Application of Equatorial Coordinates: In order to measure the equatorial coordinates, we can first run a "great circle"[8] through the north and south celestial poles (NCP, SCP), perpendicular to the equator, creating an *hour circle* at the vernal equinox, where the ecliptic intersects with the celestial equator.[9] We can then define a second hour circle through NCP, SCP, and the object of interest. The angle between the hour circles is then the right ascension, measured east from the equinox.

---

[8] A great circle is a cross-section of the surface of a sphere that is centered on the center of the sphere.
[9] Swinburne astronomy online, https://astronomy.swin.edu.au/cosmos/E/Equatorial+Coordinate+System.

When measuring the declination coordinate for a celestial object, we can use the hour circle defined above, which runs through the object and meets the equator at a right angle. The angle along this hour circle between the equator and the object is the declination (Birney et al. 2006). Therefore declination measurements are between 0° at the equator and ± 90° at the poles.

The equatorial system has two major benefits: it is a system in which stars and galaxies have a set of coordinates that does not change on an hourly or daily basis; and the Earth's rotation is aligned with the system. In practice, this means that as a star rises and sets, it can be tracked simply by moving parallel to the right ascension axis. If a telescope is mounted so that one axis of rotation is parallel to right ascension (an equatorial mount), a star can be tracked by driving along that axis at a constant rate. While the use of units of time for right ascension may seem strange, it has the advantage that a difference in right ascension translates directly into a difference in the times when those positions are at their highest point in the sky. Equatorial coordinates are thus very useful for observation planning.

Epoch and Precession: The equatorial system is tied to the Earth's rotation, but the Earth is not a perfect rotator, its rotation changing over time due to *precession* and *nutation*. The *precession of the equinoxes* is an effect by which each equinox—the points at which the Sun lies on the celestial equator—gradually shifts with respect to the background stars and galaxies, moving westwards by 50″ a year (so the Earth's alignment rotates with a 26,000 year period)[9]. Since the equatorial coordinate grid is referenced to the equinoxes, it too moves with respect to the stars and galaxies. Therefore, the equatorial coordinates of stars and galaxies slowly change.

In principle, the RA and Dec coordinates of a star or galaxy can only be correctly defined at a specific moment, or *epoch*. To address this problem, the equatorial system is "frozen" at a reference date—currently 2000.0 or the start of the year 2000 CE—and all right ascensions and declinations are given in that format. To keep the system close to the Earth's rotation, a new system is adopted every 50 years or so.

The current set of coordinates is denoted "J2000.0"[10,11]; while older measurements may be given in the "B1950.0" system.[12] The equatorial coordinates for the gravitationally lensed quasar known as QSO2237+0305 (the "Einstein Cross") were RA = 22^h 37^m, Dec = +03° 05′ for the B1950.0 system, and are now RA = 22^h 37^m, Dec = +03°21′ for the J2000.0 system[11]. Therefore, while the quasar has not changed its position, the coordinate system has slightly changed due to the effect of the Earth's precession. Because the coordinate grid simply shifts over the stars and galaxies, the effect is quite easy to remove, and online calculators that transform between these epochs (and indeed between coordinate systems) are readily available on the Internet.

---

[10] Sometimes called FK5.

[11] Swinburne astronomy online, https://astronomy.swin.edu.au/cosmos/E/Epoch.

[12] Sometimes called The Fourth Fundamental Catalog (FK4) was published in 1963, and contained 1,535 stars in various equinoxes from 1950.0.

Other Coordinate Systems

Horizontal Coordinates: The horizontal coordinate system is defined by the "plumb line" or vertical direction, which is the pole of the system, so the longitudinal plane is horizontal. This is a highly intuitive system, but the position of a star or galaxy in this system depends on the observer's position on the Earth (so it is also known as *topocentric*), and on the time of observation.

The horizon plane divides the sky into upper and lower hemispheres (Figure 9.7) —the upper hemisphere, and upper pole or *zenith*, can be seen by the observer, whereas the lower hemisphere and lower pole or *nadir* are invisible. The celestial meridian will go through the zenith, creating the north and south compass points. A great circle can be drawn from the zenith, through the position of interest, meeting the horizon at a right angle. The angular distance along this great circle from the horizon toward the zenith is the *altitude*, and the angular distance between this great circle and the celestial meridian is the *azimuth*.

Horizontal coordinates are often therefore called "alt-azimuth" coordinates, and these are the natural coordinate system for telescopes with alt-azimuth mounts. Such mounts are much easier to engineer than equatorial mounts, and so all modern large telescopes use them.

Since the horizontal system is not tied to Earth's rotation, the horizontal coordinates of a star or galaxy change from minute to minute as the Earth rotates. This makes them a poor choice to catalog or map the sky. They are far more useful as native telescope coordinates; computer systems can easily translate between

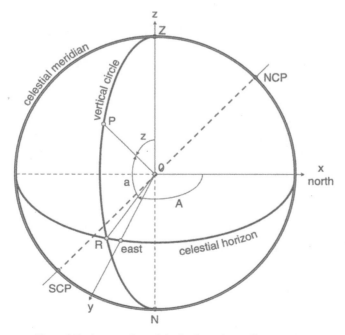

Figure 9.7. An overview of the horizontal coordinate system.

equatorial and horizontal coordinates in real time, and control the two axes of the telescope to track a position in equatorial coordinates.

Ecliptic Coordinates: Ecliptic coordinate systems use the plane of the Earth's orbit around the Sun as the plane defining the system, and can use either the Earth as a central point (for a geocentric system) or the solar system barycentre (a point inside the Sun, for a heliocentric system). As seen from the Earth, the plane of the Earth's orbit is the path of the Sun against background stars, and this is known as the ecliptic.

The relationship of the ecliptic system to the equatorial system is shown in Figure 9.8. Because the equinox is the point at which the ecliptic intersects the celestial equator, it can serve as a reference point for both systems and is used as the zero-point for ecliptic longitude (Soffel & Langhans 2012) which is measured in degrees. Ecliptic latitude is the angular distance from the ecliptic plane to the position of interest. The ecliptic poles are shifted significantly away from the celestial poles, and this is because the Earth's rotation is not completely aligned with its orbital rotation around the Sun. The angular difference (about 23°) is known as the *axial tilt* of the Earth, or the *obliquity of the ecliptic*.

Because most of the solar system is roughly coplanar, solar system objects are generally found close to the ecliptic, and move mainly in ecliptic longitude. It is therefore a natural system to use for solar system studies. The ecliptic system was predominantly used in ancient times through to several hundred years ago during the Renaissance period (Birney et al. 2006), when solar system motion was one of the chief concerns of astronomy; it is still used for solar system bodies today.

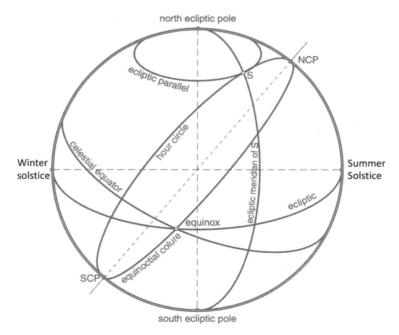

Figure 9.8. An overview of the ecliptic coordinate system, and some elements of the equatorial system.

Galactic Coordinates: Just as the solar system is a largely planar system, and so we define a set of coordinates using that plane, so the Milky Way is a largely planar set of stars orbiting the center. We can therefore define the plane of the Galaxy and use that to define Galactic coordinates, centered on the Earth. The Galactic latitude (b) is just the angular distance of the point of interest above or below the plane, and the Galactic latitude (l) is the angle along the plane (Figure 9.9).

The zero-point of Galactic longitude is taken to be the direction from the Earth to the center of the Galaxy. The Galactic center has equatorial coordinates RA = $17^h45.6^m$ and Dec = $-28°56'$ (J2000.0), lying in the constellation of Sagittarius (Birney et al. 2006).[13] The north galactic pole lies toward RA = 12^h 51.4^m and Dec = $+27°07'$ (J2000.0) in the constellation of Coma Berenices[13]. Because the structure and dynamics of the Milky Way are still an active area of research, the position of the Galactic center and pole are subject to revision; the coordinates given above are the current, revised version,[14] and may be revised again in the future. The Galactic equator is inclined by 62.9° with respect to the celestial equator (Figure 9.10).

Supergalactic Coordinates: The supergalactic coordinate system was developed by de Vaucouleurs[15] and is designed for observations of objects outside the Milky Way: galaxies, clusters of galaxies, and superclusters.[16] In this system the fundamental plane is the supergalactic plane, which is based on the "planar-like" distribution of large galactic structures in the local universe, such as the Virgo cluster, the Pisces–Perseus supercluster, and the Great Attractor about 200 million

Figure 9.9. An overview of the galactic coordinate system.

---

[13] Swinburne astronomy online, https://astronomy.swin.edu.au/cosmos/E/Epoch.

[14] Established at the International Astronomical Union (IAU) meeting in 1958.

[15] Gérard de Vaucouleurs (1918–1995), French–American astronomer.

[16] Swinburne astronomy online, https://astronomy.swin.edu.au/cosmos/S/Supergalactic+Coordinate+System.

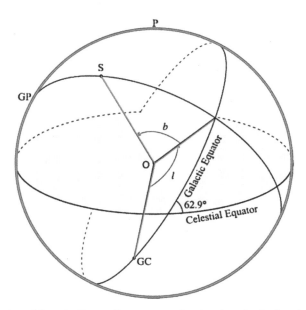

Figure 9.10. An overview of the galactic coordinate system, showing the galactic plane inclined to the celestial equator at 62.9°.

lt-yr away (the core of the Laniakea supercluster).[16] The supergalactic plane passes through the Earth in this coordinate system, since the plane is viewed from Earth. Figure 9.11 shows large scale structure in the local universe.

The supergalactic plane is roughly perpendicular to the Milky Way's galactic equator, and, when measuring supergalactic longitude (often labeled as SGL in diagrams), the zero-point is defined as the point at which the supergalactic and galactic planes intersect. This system also has established supergalactic pole coordinates, which is represented by the letters NSP in diagrams and correlates to an SGB of +90°.[16] The North Supergalactic Pole has galactic coordinates of $l = 47.37°$, $b = +6.32°$, and the zero-point of SGL has $l = 137.37°$, $b = 0°$.[16] In equatorial coordinates, the supergalactic north pole is at RA $= 18.9^h$, Dec $= +15.7°$ (J2000.0) and the zero-point is at RA $= 2.82^h$, Dec $= +59.5°$.[16]

9.2 Observation and Measurement

9.2.1 Basic Telescope Optics

In this section we discuss some of the basic operational concerns with not only optical telescopes but with telescopes of other electromagnetic wavelengths and even other messengers, such as aperture, focal ratio, magnification, field of view, and image and spectral resolution. Because most astronomers begin their career with optical telescopes (even if they are only a few years old and it is a gift from their parents), it is in optical astronomy where we first learn these terms.

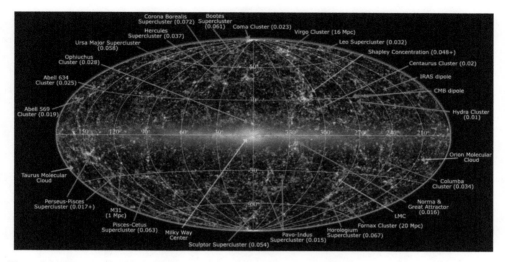

Figure 9.11. Panoramic view of the entire near-infrared sky reveals the distribution of galaxies beyond the Milky Way. The image is derived from the 2MASS Extended Source Catalog (XSC)—more than 1.5 million galaxies, and the Point Source Catalog (PSC)—nearly 0.5 billion Milky Way stars. The galaxies are color coded by redshift (numbers in parentheses) obtained from the UGC, CfA, Tully NBGC, LCRS, 2dF, 6dFGS, and SDSS surveys (and from various observations compiled by the National Aeronautics and Space Administration Extragalactic Database), or photo-metrically deduced from the K band (2.2 μm). Blue/purple are the nearest sources ($z < 0.01$), green are at moderate distances ($0.01 < z < 0.04$), and red are the most distant sources that 2MASS resolves ($0.04 < z < 0.1$). The map is projected with an equal area Aitoff in the Galactic system (Milky Way at center). Image credit: IPAC/Caltech, by Thomas Jarrett.

Aperture

Of major concern with any telescope is the size of the *aperture*. The size of the aperture determines the telescope's ability to gather light—and thus its ability to see faint objects—and the telescope's resolving power in its operating wavelengths. These are the two most important characteristics of any telescope.

The telescope's aperture is effectively equal to the diameter of its primary optical element—in almost all cases, its primary mirror. That is:

$$D = D_P = \text{Diameter of primary} \qquad (9.1)$$

and the measurement units are chosen as those most convenient for any needed calculation. However, this simple relationship is true only for an unobstructed primary. If the telescope has a secondary element which obstructs some of the incoming light in the path of the primary, the effective aperture becomes:

$$D = [D_P^2 - D_S^2]^{1/2}. \qquad (9.2)$$

We are ignoring optical effects, such as diffraction, which play a role in the more accurate determination of the effective aperture. The equations given here are close enough for practical use.

Focal Ratio
In photography, the *focal ratio* is a description of the "speed" of a lens, which is a description of the exposure time for a photograph. The lower the focal ratio (or $f/\#$), the lower the exposure time and the "faster the lens." It plays a similar role in telescopes (see below). Focal ratio also affects the depth of field of the camera, but there is no such consideration for astronomical telescopes, because all sources are at optical infinity.

If the primary mirror of a telescope has diameter or aperture D, and the distance from the mirror surface to the prime focus is l, the focal ratio of the primary mirror is:

$$f = l/D. \qquad (9.3)$$

If the telescope works only at prime focus, this will also be the focal ratio of the telescope, but more often, other mirrors are added to produce e.g., a Cassegrain focus, and each focus of the telescope will have its own focal ratio. Figure 9.12 shows the concept of the calculation. The focal length of the telescope can be substantially longer than the length of the telescope tube.

Magnification
Magnification is probably the least of the concerns with a research telescope, except for its relationship to the telescope's field of view. In the case of a telescope with an eyepiece, the magnification is the ratio of the focal length of the telescope l, to the focal length of the eyepiece l_e:

$$M = \frac{l}{l_e}. \qquad (9.4)$$

This is not relevant to a telescope which delivers its light into an instrument—camera, spectrograph, etc—rather than an eyepiece. In this case the *exit pupil*, e, is defined as the diameter of the light beam exiting the telescope, and the magnification is:

$$M = \frac{D}{e}. \qquad (9.5)$$

Field of View and Plate Scale
The *field of view* is the area of the sky that can be seen through the telescope. It is sometimes stated in solid angle dimensions such as square degrees or square arcminutes/arcseconds. For a telescope with an eyepiece, the field of view is usually

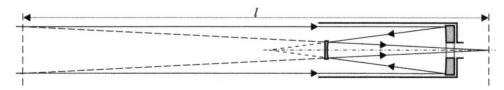

Figure 9.12. The focal length of a multiple mirror focus system telescope.

given as the diameter of the circular field. This is determined by the eyepiece design and the magnification of the telescope-eyepiece system. For example, if an eyepiece has a natural field of view of 40° and is placed in a telescope such that there is a magnification of 40, then the circular field of view becomes one degree across.

When an imaging technology is used, the imaging surface intercepts the light from the telescope, and an image is projected onto it. This imaging surface could be a photographic plate, a charge-coupled device (CCD) or CMOS camera array (Section 9.2.4), or even the ends of optical fibers to pick off the light for further analysis. The relationship of linear size on the imaging surface to angular size on the sky is given by the *plate scale*. The plate scale can be measured by allowing the light of two stars to enter the telescope aperture, one star on the telescope's optical axis, and the other slightly off-axis so the star's light path forms an angle α with the axis. On the image plane at the focus of the telescope, the image of the second star will be at distance s from the center, which is the image location of the on-axis star. If the focal length of the telescope is l:

$$s = l \tan \alpha. \tag{9.6}$$

Because α is usually a very small angle, we can write $s \approx l\alpha$, where α is measured in radians. The plate scale

$$P = \frac{\alpha}{s} = \frac{1}{l} \tag{9.7}$$

with units of radians per meter, or just m^{-1}. In practice, the plate scale is given in arcseconds per millimeter. For example, a two meter, $f/16$ telescope creates a plate scale of 0.031 rad m^{-1} which is equivalent to 6.5″ mm^{-1}. If the imaging device is 25 mm in size, then the field of view becomes 25 mm ×6.5″ mm^{-1} = 160″ = 2.7′.

Image Energy
In order to detect light, there must be sufficient energy available to affect the detector, so the amount of energy from an astrophysical source that falls onto the imaging surface determines detectability.

In the case of a compact source imaged in such a way that all the light from it falls onto a single detecting element of the imaging surface (e.g., a pixel for a CCD), the energy gathered by that element depends only on the telescope aperture and the exposure or integration time—the time over which the energy is measured. In general, however, the light from one source will be spread over multiple detecting elements. For a circular object with angular size α producing an image of diameter s, the area of the image is proportional to s^2; the incoming energy (proportional to D^2) is therefore spread over a number of pixels proportional to s^2, and the energy per pixel per unit time is proportional to s^{-2}, and therefore proportional to l^{-2}: Thus, the energy supplied (per unit time) to each pixel is:

$$E_P \propto \left[\frac{D}{l}\right]^2 \tag{9.8}$$

which is just inversely proportional to the square of the focal ratio. So the greater the focal ratio, the less energy is accumulated per unit area; in order to detect an object, this must be compensated for by a longer integration time, so a high-focal-ratio system is "slow." A greater focal ratio means higher resolution but lower energy per unit time supplied to each pixel.

Image Resolution
The spatial resolving power of the telescope is governed by the aperture and the observed wavelength.[17] The physical limit of resolution is due to the diffraction of light as it enters the telescope. A circular aperture (such as the mirror surface) is a two-dimensional version of a single-slit; rather than a set of fringes, its diffraction pattern is circular spot surrounded by a set of concentric rings. George Airy derived the distribution that describes this pattern, still known as the Airy function. (Section 5.2.3 and Appendix E).

Two point sources of light (such as stars) can be resolved if their Airy functions can be distinguished from one another. Rayleigh suggested a criterion to reflect this: if the center of the Airy disk of one star lies on the first dark fringe (or edge) of the Airy disk of a second, then the two stars are said to be resolved. The angular resolution α (i.e., the angular separation of two stars that satisfy the Rayleigh criterion) is given by:

$$\sin \alpha = 1.220 \, \frac{\lambda}{D} \tag{9.9}$$

where λ is the wavelength of the observed radiation and D is the diameter of the aperture.

The spatial resolution Δs can also be calculated as a physical length on the imaging surface, using the focal length l:

$$\Delta s = l \tan \alpha \approx l \sin \alpha \approx 1.220 \, \frac{\lambda l}{D} \approx 1.220 \, f\lambda \tag{9.10}$$

where f is the telescope's focal ratio.

For a telescope with a focal ratio of $f/15$, observing green light ($\lambda = 550$ nm), the angular resolution is 4.5×10^{-8} rad or 9×10^{-3} and the linear resolution is 10 μm. This is the physical limit of resolution—the actual resolution attained will depend on many other factors, most notably atmospheric seeing, which generally limits resolution to about an arcsecond.

Stray Light
A perfect telescope collects the light from a small solid angle of the sky and concentrates it in a well-defined way onto an imaging surface. Additional light, or light outside the defined pattern, is stray light. This can have two effects: if it is spread out over the imaging surface, it will simply increase the background, reducing

---

[17] In the optical region, the standard wavelength for calculating resolution is that of green light, 550 nm.

the sensitivity of the system. If the stray light falls onto a small area of the imaging surface, it may look like a star or galaxy, creating a spurious object detection.

Stray light generally comes from bright light sources that are fairly close to the line of sight of the telescope.[18] This could be the Moon or a planet; or it could simply be a bright star nearby. If the light is reflected onto the imaging surface by *specular reflection* (in which a beam of light is reflected at a specific angle, and so remains concentrated), a spurious source may be seen. If the reflections happen off rough surfaces, the light will be reflected over a broad range of angles, and will become a diffuse background.

The most important element to controlling stray light is to prevent specular reflections. These could be from elements of the telescope structure itself, from the local environment (e.g., a telescope dome), or even internal reflections from optical elements. If these surfaces are made rough, they will not produce large specular reflections, and instead increase the background. If possible, surfaces should be made non-reflective, so that stray light is generally absorbed rather than reflected. Careful optical design can also minimize the problem.

Spurious sources due to stray light can be identified by moving the telescope slightly. As the field of view shifts, the spurious sources will generally move differently to the star field.

9.2.2 Astrometry

Astrometry is the study of the positions of stars and galaxies in the sky. Its first aim is to provide an accurate and reproducible framework by which any position in the sky can be converted to equatorial coordinates. This framework—the International Celestial Reference System or ICRS—is currently based on radio interferometric measurements of a sample of extragalactic radio sources, and other catalogs' measurements of object coordinates are tied to this system.

This system of sources with well-defined coordinates allows a general conversion to be established for an astronomical image between the position in the image and celestial coordinates. The position on the image will generally be measured as x- and y-values, by physical measurement of a photographic plate or, more likely, as pixel values read off a digital image. If the x-axis is aligned with right ascension α and the y-axis is aligned with declination δ, the transformation from (x, y) to (α, δ) is:

$$\alpha = \alpha_0 + \alpha_p(x - x_0) \quad \delta = \delta_0 + \delta_p(y - y_0) \tag{9.11}$$

where (x_0, y_0) and (α_0, δ_0) refer to a fiducial point with known coordinates, and α_p and δ_p are the plate scales or pixel scales of the imaging system. These scales are usually equal in magnitude, but α_p is usually negative (since right ascension conventionally increases to the left of an astronomical image, but the image position is conventionally measured to the right), while δ_p is positive. In the general case, where x- and y-axes are not perfectly aligned to α and δ, each coordinate will depend on both x and y, defined by a *position angle* of the field, usually measured in degrees

---

[18] But not always; stray light has been seen from car headlights near an observatory.

east of north on the sky. In addition to this field rotation, more complex terms may be added to the transformation to take account of distortions of the image.

While most astrometry is carried out simply to define the positions of observed objects, very precise astrometric measurements can be used to measure small shifts in the positions of objects, allowing for measurements of distance by parallax, and *proper motion.*

9.2.3 Photometry

Photometry is the measure of the intensity of light (or more generally, the flux of radiation) from an object. This is one of the most fundamental measurements that we can make in astronomy. Section 3.1.2 covers the magnitude and color systems that are used to present these data. Here, we consider the methods used to count the photons or measure the energy from the object of interest. In photometry, we are not interested in imaging, but in counting photons; however, most modern photometry is performed with imaging detectors such as CCDs.

Photometers

Specialist *photometers* are instruments that measure the photon flux coming into them, without any attempt at imaging. One common type uses a *photomultiplier tube* (PMT; Figure 9.13) to detect and count the photons, generating a current pulse for each photon arriving. The tube has a cathode and an anode with a series of equally spaced dynodes between them. An electric potential of 2000 to 4000 V is applied with a resistor ladder creating stepped potentials on the dynodes. In some versions the passive resistor ladder is replaced by active components, which decreases the response time of the PMT. A photon enters the PMT and strikes the photocathode, knocking off an electron. This electron is attracted to, and strikes the first dynode, knocking off more electrons, which hit the succeeding dynodes creating a cascade of electrons. This creates a current spike at the anode, which is detected by the sensing electronics, and the photon is counted.

The instrument operates by passing the light from the telescope through the photometer's aperture and through a filter which selects the passband (e.g., *UBV*

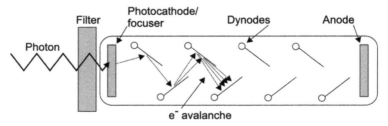

Figure 9.13. A schematic of the operation of a photomultiplier tube (PMT) used as a photometer. Photomultipliers are generally sensitive to ultraviolet, optical, and near-infrared. A photon enters the tube and liberates an electron from the photocathode. This electron liberates more electrons from the succeeding dynodes, eventually creating a current flow which can be detected by the electronics. Avalanche photodiodes are the solid-state equivalent of vacuum photomultiplier tubes, but these are not used very often in astronomy.

filters, Section 3.1.2) for the measurement. The sensitivity of the photocathode is not uniform across its surface. To obtain the most consistent readings, we must keep the starlight at the same position on the face of the photocathode. Atmospheric turbulence, tracking errors and other effects naturally cause the telescope's image of the star to drift. A Fabry lens[19] is used to effectively eliminate this drift by removing any positional information about the source. The Fabry lens defocuses the star's image, spreading it out on the photocathode's surface. The star can vary in its position in the aperture and the illumination of the photocathode remains consistent.

9.2.4 Astronomical Imaging

The original astronomical imaging system was the observer's eye. This naturally led to the hand, with pencil and paper. Next came photographic plates and films, which have similar frequency response and sensitivity to the eye, but produce a lasting record in a reproducible way. This was the only imaging technology available for about a century,[20] and was the dominant technology for even longer (Ford 2014). Digital imaging arrays have now supplanted these older methods, producing data with high resolution and dynamic range that are immediately available to the observer (since the chemical processing of photographic emulsions is not required). As shown in Figure 9.14, a digital image is simply an array of picture elements, called *pixels*, which are arranged in rows and columns; the value of each pixel is a

Figure 9.14. An example of a digital image showing the matrix of its building blocks called picture elements or pixels. Image credit: ESA/Hubble & NASA http://www.spacetelescope.org/projects/fits_liberator/improc/.

---

[19] Named for French physicist, Charles Fabry (1867–1945).
[20] The first known successful astrophotograph was taken by John Draper (1811–1882) in March 1840, who pictured the Moon on a photographic plate with a 20-min exposure.

measurement of the amount of light that fell on the corresponding detector element in the array.

The main visible-light imaging technology in use today is the *charge-coupled device (CCD)*, developed in 1969 at AT&T Bell Labs by Willard S. Boyle (1924–2011) and George E. Smith (1930–). The CCD uses the fact that semiconductors—doped silicon in particular—can be used to make integrated circuits and also can be *photoconductors*, in which light excites electrons from the lower-energy valence band to the higher-energy conduction band. By measuring the charge built up in this way, the brightness of the light can be measured. It is therefore possible to build a solid-state device that incorporates both the light-detecting sensor material and the processing circuitry. The first, commercially available CCD came out in 1974, with dimensions of 100×100 *pixels* (picture elements). Current CCDs have pixel dimensions in the thousands and pixel counts in the millions (Figure 9.15).

A CCD is a microchip onto which the light collected by the telescope is focused, consisting of a grid of individual light-sensing elements called metal-oxide semiconductor (MOS) capacitors. Each of these capacitors is wafered on a layer of silicon. Significant advantages of modern CCDs include their *quantum efficiency*[21] (in excess of 70%), dynamic range (Ford 2014), linearity, and almost-immediate data feedback to the observer.

CCDs have been the most common visible and near-ultraviolet (~300–400 nm) imaging sensors in astronomy. The continued development in CCD technology from the 1980s to the present led to a significant increase in its sensitivity to wavelengths

Figure 9.15. A 2.1 megapixel CCD chip from an Argus brand digital camera, typical of commercial CCDs. Image credit: Wikipedia: Merzperson https://en.wikipedia.org/wiki/File:ArgusCCD.jpg.

---

[21] The quantum efficiency of a CCD is a property of the photovoltaic response defined as the number of electron–hole pairs created and successfully read out by the device for each incoming photon.

from gamma and X-rays to near-infrared. Infrared and optical photons have enough energy to produce one electron per photon in the CCD's light-sensing elements, allowing a direct count of incoming photons with a linear relationship between the signal and the light intensity. Unfortunately they are non-linearly dependent on frequency. Photons with too low energy (frequency) cannot penetrate the CCD material and those with too high energy liberate electrons too deep in the substrate to allow collecting the energy of ultraviolet or X-ray photons; tens to thousands of electrons may be produced in each pixel, possibly reaching the limit of sensing the element's electron capacity. However if the capacity is not exceeded, the number of electrons produced is still proportional to the photon's energy and the CCD can still readout the correct count and energy of the incident photons; luckily, the number of incoming X-ray photons is very low, even from the strongest celestial sources. A rule of thumb for the MOS capacitor's electron capacity is to take the size of the pixel in micrometers and multiply by one thousand.

Cooling a CCD reduces the device's dark current, improving the sensitivity of the CCD to low-light intensities. Astronomical CCDs are generally cooled well below freezing to give the best performance.

CCD Structure and Operation
CCDs are built with several different architectures. The most common are full-frame, frame-transfer, and interline. The difference between these architectures is the method of shuttering and of transporting the collected charge. Almost all astro-nomical instruments (using a CCD detector) use a full-frame detector, so we will limit our discussion to this architecture (other architectures are sometimes used for high-speed imaging). This architecture is the easiest to fabricate and operate, and it works well for the needs of astronomers mostly because this architecture renders the highest resolution image.

The elementary light-sensing component of a CCD is the pixel (Figure 9.16), made from MOS capacitors, also called photodiodes. Millions of these MOS capacitors are arranged into a square or rectangular array of columns and rows, the surface of which defines the imaging surface of the device. The individual pixels are created in one dimension (as columns) by the polysilicon[22] gate electrode strips and in the other dimension (as rows) by electrical barriers with the silicon substrate.

With a full-frame CCD, a mechanical shutter must be used in front of the CCD device or the image will smear as the pixel electron charge is read out. The first stage of the process of building an image is the collection of photons and the trans-formation of those photons to collected electrons. This is the job of the capacitor (pixel). The number of collected electrons is proportional to the number of collected photons, which is proportional to the intensity of the source and the exposure time. When the shutter is closed, the electrons in each capacitor must be counted. The method of unloading and counting the electrons from the pixels is where the CCD

---

[22] Polycrystalline silicon (polysilicon) is 99.9% pure silicon.

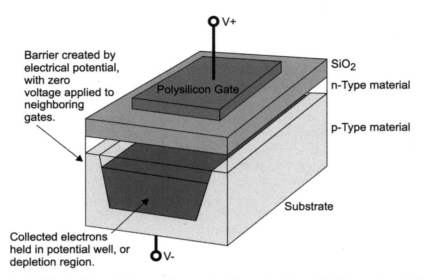

Figure 9.16. A CCD pixel is made from a silicon metal-oxide semiconductor (MOS) capacitor, which stores electrons liberated from the silicon substrate by incoming photons. The electrons are stored within a potential well (the p–n junction depletion region) created by placing a small voltage across the gate to the substrate. To create the well, the neighboring gates are held at a neutral or zero voltage.

gets its name—the charge from each pixel is transferred to the next, and eventually read out when the charge reaches the edge of the array.

The circuitry used to move the charge on the array depends on the precise architecture: Four-Phase (4ϕ), Three-Phase (3ϕ), Two-Phase (2ϕ), and Virtual-Phase (Vϕ) are used, with advantages and disadvantages to each.

Charge Readout: At the output end of the row shift register there is a "floating diffuse sense node." This sense node is drained of charge and the potential in this state is measured and used as the zero voltage reference. A pixel's charge is then pushed into this sense node and the change in potential within the node is measured by a source follower type amplifier. The new potential is converted to a voltage difference (from the reference) and this is sent off-chip for conversion to a digital value. The diffuse sense node is drained and the next pixel charge is deposited for measurement.

Other Array Technologies

CCD-like devices have also been made with other technologies. For visible-light imaging, CMOS sensors can be used. These are similar to CCDs, but use different processes. They do not transfer charge between pixels, but each pixel is read out individually. They can generally be read out much faster than CCDs, and are cheaper to manufacture, but their performance is generally inferior. Increasingly, however, they are being used as alternatives to CCDs.

Because CCDs are built of silicon, their wavelength sensitivity is restricted to that of silicon photoconduction, which is insensitive to infrared light past about 1000 nm wavelength. In order to make an array detector for near-infrared light, the photoconductor elements must be made of a specialized material (e.g., indium antimonide, InSb or mercury cadmium telluride, HgCdTe) which is then bonded to a silicon backing containing all the circuitry.

9.2.5 Spectroscopy

Spectroscopy, for any messenger, is the business of analyzing that messenger by the energy contained. In the case of electromagnetic radiation, this means analyzing by frequency or wavelength. While color filters can be used as a form of coarse spectroscopy, the term is usually used to refer to analysis by *dispersion* of the light, so that the path of light with one wavelength will be different to the path of light with another wavelength. The instruments that do this are generally known as *spectrographs*. In the optical and near-optical regime,[23] they generally do this by picking the light from a specific location in the focal plane of the telescope (where an imaging surface would otherwise be placed), dispersing the light with a prism or, more often, a grating, and reimaging the dispersed light onto an imaging detector that will measure the light at various wavelengths.

Optics Before Dispersion
The simplest way to deliver light from the telescope to the dispersion element is to pass it through a narrow slit, whose narrow dimension is parallel to the direction in which the light will be dispersed. The light coming through the slit is imaged and dispersed, leading to a series of slit images stacked along the dispersion direction—a spectrum.

Because only the light coming through the slit has been allowed into the spectrograph, only a small part of the telescope field can be observed. For large-scale spectroscopic surveys of many objects, observing multiple objects in the same field of view provides a compelling advantage. This can be done by replacing the slit with a "lenslet" on the end of an optical fiber. Many lenslets can then pick off the light from many sources, and transport it into the spectrograph. Integral field spectroscopy is a development of multiobject spectroscopy: If the lenslets are bunched together, then a spectrum can be taken toward many contiguous regions in an image, providing a form of imaging spectroscopy.

Beam Splitting: The incoming light can also be divided by wavelength, so that shorter-wavelength (blue) light and longer-wavelength (red) light are split into two separate beams, each one of which can then be dispersed and imaged. This beam splitting is usually done with a *dichroic filter* (Figure 9.17)—a filter that allows either red or blue light through, and reflects the rest, producing a transmitted and a reflected beam. These filters can be a simple metallic coating, but are more often carefully tuned, using interference between reflected and transmitted waves created by layers of optical coatings vacuum deposited onto a glass base. The coatings then select the frequency point at which light is reflected by the filter rather than transmitted. A dichroic is basically a series of *Fabry–Pérot interferometers*.[24]

---

[23] X-ray spectroscopy is described in the X-ray chapter of Book 2.
[24] Named for French physicists, Charles Fabry (1867–1945) and Alfred Pérot (1863–1925).

Figure 9.17. Dichroic filters are used to split an incoming light beam into a "red beam" and "blue beam" in a dual-beam spectrograph. Image credit: NASA.

Prisms and Diffraction Gratings

There are a few different optical components that can disperse light according to wavelength:

1. Prisms refract light through an angle that depends on the wavelength;
2. Diffraction gratings (technically *transmission diffraction gratings*) are made up of alternating opaque and transparent stripes. The transparent stripes act as slits, allowing the light through, some of it in a straight line and some of it at an angle that depends on the wavelength;
3. Reflection gratings (technically *reflection diffraction gratings*) are similar to transmission gratings, but have reflective surfaces interrupted by regular lines that reflect some of the light at the normal angle of reflection and some at an increased angle that depends on wavelength;
4. Grisms combine gratings with prisms, so that the refraction in the prism cancels out the change in angle due to the grating, and the beam passes through the grism along a straight line, but still dispersed.

While the action of a prism was known since Isaac Newton's work, diffraction gratings were developed at the end of the 19th century, requiring major technological advances. Early gratings were mechanically diamond-cut mirrors, as produced by Grayson[25] in Australia and Rowland[26] in the USA. Modern gratings

---

[25] Henry Joseph Grayson (1856–1918), British-Australian scientist.
[26] Henry Augustus Rowland (1848–1901), professor of physics at Johns Hopkins University.

often have optically-produced patterns, where the pattern itself is generated by interference fringes from lasers, and this is photographically transformed into a grating, either reflection or transmission—these are known as holographic gratings. In practice, gratings are used with line densities of hundreds to thousands of lines per millimeter (mm^{-1}). The reciprocal of the line density is the line period or line spacing in mm, which defines the dispersion of the light.

While a prism simply refracts and spreads out the incoming beam of light into an outgoing spectrum, a grating diffracts incoming light by multiple discrete angles. When a light beam of wavelength λ is incident on a reflection grating with angle θ_i (measured from the grating normal), it is reflected from the grating at a series of angles $\theta_m(\lambda)$. If θ_m is measured from the normal in the same sense as θ_i it will generally be negative, while θ_i is positive (Figure 9.18). The reflected light is diffracted into several spectral images or *orders*, defined by integer values $m = 0, \pm1, \pm2...$, which in turn define θ_m. The grating equation for a reflection grating is:

$$d(\sin \theta_m (\lambda) + \sin \theta_i) = m\,\lambda. \qquad (9.12)$$

The line period d must be in the same length units (e.g., mm) as λ. For $m = 0$, $\theta_m = -\theta_i$, giving specular reflection for all wavelengths. For non-zero m, the beams diverge from specular reflection, traveling closer to the plane of the grating with increasing positive order and increasing wavelength, and closer to the normal to the grating with increasing negative order and negative wavelength. Consecutive orders may overlap: long-wavelength diffraction spots in one order may be found near short-wavelength diffraction spots in the next order. The higher the spectral order, the greater the overlap into the next order.

Transmission gratings obey the same equation, except that the zero order corresponds to straight-line propagation rather than specular reflection. $\theta_m(\lambda)$ is then the angle by which the beam of order m is deviated from the straight line:

$$d \sin \theta_m(\lambda) = m\,\lambda. \qquad (9.13)$$

Blazing: The light that falls onto a diffraction grating is distributed over all the orders according to a single-slit diffraction pattern corresponding to the width of the reflecting or transmitting elements of the grating. Because a single-slit diffraction pattern peaks at the center and rolls off to the sides, the most light goes into the $m = 0$ spot, with decreasing amounts of light in higher orders (which also have

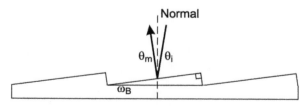

Figure 9.18. A reflection grating in cross-section. This grating is blazed, with the grooves cut in a triangular faceted shape, meeting the Littrow condition.

higher dispersion). A grating spectrograph thus tends to waste light, which makes it inefficient.

A way around this problem is to change the cross-section of the grating; this is mainly done with reflection gratings, and is called *blazing*. The cross-section consists of triangular facets, so that the light reflects off surfaces that are at an angle (the *blaze angle*, ω_B) to the plane of the grating (Figure 9.18). The diffraction pattern still peaks normal to the grating itself, but the specular reflection ($m = 0$) peak has now shifted by ω_B away from the peak of the diffraction pattern, and the diffracted orders have shifted toward the peak. By tuning the blaze angle and the optical configuration, a desired diffraction order can be moved to the single-slit diffraction peak, giving a much higher efficiency. The blazing performance is wavelength-dependent, so a design wavelength must be specified.

In practice, most blazed gratings are used in the Littrow[27] configuration, where the light is reflected back in the same direction as the incoming light ($\theta_m = \theta_i = \omega_B$).

Spectrographs

Spectral Resolution: If a spectroscopic instrument can distinguish between two spectral lines separated in wavelength by $\delta\lambda$, but not by less (in the same way that an imaging system can distinguish between two stars), then $\delta\lambda$ is the spectral resolution. It can be given as a fractional quantity $\delta\lambda/\lambda$, where λ is the average wavelength of the two lines being resolved. The reciprocal of this fractional resolution is the *resolving power*, $\mathcal{R}$, of the spectrograph. In the case of a grating spectrometer:

$$\mathcal{R} = \frac{\lambda}{\delta\lambda} = m\,N \qquad (9.14)$$

where m is the diffraction order number and N is the total number of lines in the diffraction grating. For example, a diffraction grating with $n = 300$ mm^{-1} and width 100 mm has a theoretical resolution limit at $\lambda = 5500$ Å and order $m = 1$ of $\delta\lambda = 5500/(1 \times 300 \times 100) = 0.18$Å. Low-resolution spectroscopy is generally considered to use $\mathcal{R} < 1000$, high-resolution spectroscopy uses $\mathcal{R} > 10,000$, and medium resolution would have $\mathcal{R}$ in the thousands.

A full spectroscopic system consists of the telescope that gathers the light and delivers it to the spectrograph, the spectrograph itself, and the imaging system that captures and measures the dispersed light. This is a complicated system, and its analysis is out of our scope. In general, however, the spectral resolving power of the system will depend on more factors than just the theoretical formula given above.

The spectrograph entrance slit is imaged to form the spectrum, giving an infinite series of slit images, so the resolution cannot be finer than the slit width. A wider slit decreases the resolution, but will allow more light in, improving sensitivity. For ground-based telescopes, seeing becomes important, since the light will be spread out over the seeing disk, and so the slit must be widened to take account of that, with the accompanying loss of resolving power.

---

[27] Joseph von Littrow (1781–1840), Austrian astronomer.

Echelle Gratings: As spectral resolution increases, it is easy for the length of the spectrum to exceed the dimensions of the detector. Imaging a very high-resolution spectrum on a square detector array can be done with an *échelle*[28] spectrograph. This uses a blazed grating, but the incoming light has a high angle of incidence, so that it reflects off the short side of the triangular facets (Figure 9.18), and the grating looks like a flight of stairs. The greatly increased blaze angle allows the use of high-order diffraction ($m \sim 40 - 150$), where a number of spectral orders overlap each other. A second, low-resolution diffraction grating, oriented perpendicular to the échelle line pattern is then used to *cross-disperse* this spectrum, splitting the different orders into a stack of spectral segments on the detector (see Figure 9.19). This arrangement can provide resolving powers of order 100,000, and échelle spectrographs are widely used e.g., for radial velocity detection of extrasolar planets.

Fabry–Pérot Spectrograph: Fabry–Pérot Spectrographs do not use dispersive optics such as gratings. Instead, they use Fabry–Pérot interferometers as very narrowband tunable filters. Two highly reflective surfaces are placed very close to each other in parallel and placed in a collimated telescope beam (Figure 9.20). Usually, the pair of surfaces are together called the *étalon* but technically speaking the first (incoming light) surface is the étalon and the second (outgoing light) surface is the interferometer. Multiple reflections between the two surfaces create a destructive interference pattern, except for one wavelength, which depends on the gap between the plates and the angle of the incoming beam.

The light emerges in a circular pattern that is focused onto the detector face. The spectral width is very narrow and the central wavelength is changed by adjusting the surface gap. This type of spectrograph provides high resolutions of $\mathcal{R} \approx 10^4$ or

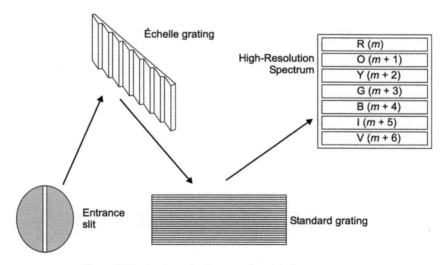

Figure 9.19. A schematic diagram of an échelle spectrometer.

---

[28] The French word for stairs or ladder.

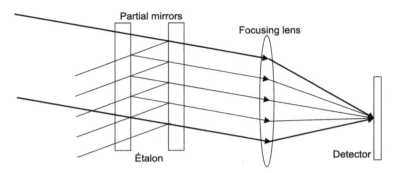

Figure 9.20. In a Fabry–Pérot spectrograph, the étalon creates an interference pattern that allows only a very narrow passband. The central wavelength is controlled by the gap in the étalon and the incident angle of the incoming beam. A circular interference pattern is created on the face of the detector.

higher. Because their spectral width is so narrow, Fabry–Pérot spectrographs are used mainly to study spectral line profiles, and because they only allow through one wavelength at a time, a spectrum has to be built up by observing one wavelength after another, so they are not well-suited to broadband spectroscopy.

Fourier Transform Spectrometer: The *Fourier transform spectrometer*, also non-dispersive, uses a Michelson interferometer to create a spectrum. The incoming beam is collimated and split into two paths at 90° to each other. These two beams are then reflected back by mirrors, one fixed, the other movable (the standard Michelson interferometer design). The two reflected beams create an interference pattern at the surface of the detector.

For a monochromatic source, the movable mirror creates successive constructive and destructive interference leading to a variation in the intensity of the light at the detector. The intensity varies as the cosine of the distance moved by the mirror, s, as $\cos(4\pi s/\lambda)$. For the light from a telescope, the intensity is the sum over the wavelengths of these cosine terms and the spectrum is extracted by an inverse Fourier transform.

Similar to the Fabry–Pérot spectrograph, the Fourier transform spectrograph can reach $\mathcal{R} \approx 10^5$ but with a much wider passband. The main disadvantage is the signal to noise ratio. The photons from the entire spectrum allowed through the passband contribute to the noise while only those photons analyzed (by mirror movement) contribute to the signal.

9.2.6 Polarimetry

The polarization state of the incoming light can provide critical information about the environment of the source such as the structure of magnetic fields. This adds an important data set to those of spectroscopy, imaging and photometry. The majority of data gathered from celestial sources is concerned with the thermal properties of the objects. This means the objects are static—in hydrostatic equilibrium or local thermal equilibrium. However, most objects are evolving or dynamically changing with time, sometimes involving very violent events (like supernovae).

The emissions from celestial objects can be polarized by a number of mechanisms. It may be polarized while in transit (magnetic fields or various scattering mechanisms) or by the dynamics of the system (asymmetrical explosions, rotational or orbital motions). It may also be intrinsically polarized by the mechanism of emission (maser emissions, synchrotron emissions, plasma oscillations). Thus *polarimetry*, the measurement of the polarization of light, provides valuable data on the dynamics of emission sources.

Because polarimetry can discern the geometry of dynamical environments, it is used for classifying various, related objects such as active galactic nuclei. Polarimetry provides important data about the dynamics of stellar and pulsar winds which enrich the interstellar medium. The study of the magnetic fields of normal stars, white dwarfs, pulsars, and spectroscopic binaries with polarimetry gives us deep insight into the dynamics (such as the angle of inclination of the orbital plane of spectroscopic binaries) of these systems. Polarimetry also provides important data on the character of the interstellar medium, including the geometry and alignment of dust grains, and thus provides information important for understanding the formation of stars. Non-symmetrical dust grains tend toward physical alignment due to the galactic magnetic field. This alignment (with a scattering probability depending on grain size) causes a slight polarization of starlight and thus an indication of the geometry of the galaxy's magnetic field.

The measurement of the polarization of light is not simple. Polarization data is generally very close to the signal noise level. The exposure or integration time must often be 100 to 1000 times longer than those required for spectroscopy to get a usable signal to noise ratio. The instrument itself causes polarization of the light. Optical coatings (the mirror coating may not be uniform) or non-normal reflections (such as tertiary or folding mirrors[29]) within the telescope can also cause additional polarization. This can be removed from the data by using a calibration source. This also requires a very stable platform for the instrument so the internal polarization can be removed from the data. Even with this stable platform, long-term tracking causes the instrument polarization to rotate on the image. For space-based observation (the most useful, particularly in the UV range) stabilization can be a difficult task. Non-linearities in the light detection device (usually a CCD) also add polarizing effects to the data. Polarized and polarizing background emissions must also be taken into consideration during data reduction. Also the dynamics of the source usually cause some variability in the polarization and thus observations must be repeated in order to gain enough data to create a useful understanding of the system.

The basic construction of a common type of polarimeter is shown in Figure 9.21. A *retarder plate*, or wave plate, is placed far upstream in the optical path. A retarder plate is made from a birefringent crystal (such as calcite) with a specific thickness. These crystals have an ordinary (light is polarized perpendicular to the axis of anisotropy) and an extraordinary (light is polarized parallel to the axis of anisotropy) polarization axis. The crystal is carefully cut with the extraordinary

---

[29] This means that polarimeters cannot be placed at the coudé or Nasmyth foci of telescopes because these require the use of folding mirrors.

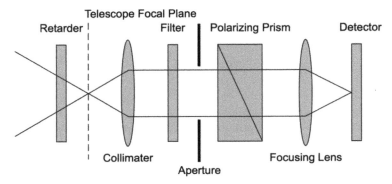

Figure 9.21. A schematic overview of a polarimeter.

axis parallel to the surfaces of the plate. When the extraordinary axis refractive index is smaller than the ordinary axis index, the extraordinary axis is called the "fast axis" and the ordinary axis is called the "slow axis." Light polarized along the fast axis propagates faster than light polarized along the slow axis. Thus, depending on the thickness of the crystal, light with polarization components along both axes will emerge in a different polarization state. A *half-wave plate* causes the light from the two axes to be shifted by 180° relative to each other, while a *quarter-wave plate* introduces a 90° shift. The retarder plate is usually rotatable. The light from the retarder is then passed through an analyzer, most often a *Wollaston prism*, which separates the rays of different polarization states into slightly different paths. These light rays then travel to the detector, where they can be imaged simultaneously. Comparison of the images that come through the different pathways can then be used to calculate the polarization of the source.

9.3 Data Storage and Access

The earliest form of astronomical data storage was the catalog of stars or galaxies, which preserved the record of observations in written form. As photography became a key technology, the archiving of photographic plates became a new requirement. Almost without exception, modern astronomical data is recorded electronically, and stored digitally. The digital nature of modern data makes it easily accessible through computer networks, but its scale can prove a significant obstacle to access.

9.3.1 Data Storage

The majority of astronomical data is in the form of multidimensional arrays or lists of numbers; each number represents the same kind of measurement performed at a different point in space, time, frequency, etc. Data storage formats tend to follow this basic pattern, and have the following functions:

1. Storage of the actual data numbers;
2. definition of the data structure, so that any given data number can be associated with its position in the multidimensional data space;

3. Documentation of the data taking and processing (*metadata*) to allow further analysis and interpretation of the data.

Formats that include all these functions give robust self-documenting data storage with minimal barriers to access. Over the late 20th century, many different data formats were developed and used, but gradually the *Flexible Image Transport System* or FITS became the dominant astronomical data storage format.

FITS: The FITS format was developed in the 1970s and first published as a standard in 1981 (Wells et al. 1981). In this initial form, defined for two-dimensional images, it consists of a list of keyword-value pair records (the *header*) that store the definitions and metadata, followed by the data themselves. In the decades since, it has expanded to allow multidimensional data storage, the storage of multiple header-data units in one file, and the storage of data in tabular form. It has remained fully backwards-compatible, so that software to handle modern FITS files will read and manipulate FITS files from the 1980s without error.

Several advantages can be recognized in its design:

- The header is stored in ASCII format, and so is instantly machine-readable;
- the list of key-value pairs is unlimited, and can be expanded as necessary;
- data and metadata are stored in the same file, so metadata will never be disassociated from its data.

Overview of Astronomical Software

Astronomical software has evolved very quickly through the last few decades. At the end of the 20th century, there were many software packages available, ranging from small programmes dedicated to specific instruments or analysis techniques to large software suites that covered many data sources and techniques. These were generally designed for scientist end-users who were assumed to have little knowledge of programming, and written in fairly low-level languages such as FORTRAN or C.

Many of these larger suites included their own high-level programming languages to allow the user to automate data handling and analysis. This functionality came together with higher-level languages to generate a new software ecosystem, in which most astronomical data handling and analysis is handled by programmes, functions, and libraries that are glued together by high-level interpreted computer languages, mainly python. An implication of this paradigm is that scientists must be sufficiently comfortable with programming that they can write their own procedures.

9.3.2 Data Access

Data Archives

The vast majority of astronomical data is archived, often in multiple locations. Most observatories operate their own local archives, and there are many additional national-level archives. Most astronomical data is stored in a few well-known formats (usually FITS) and most observatories follow a strict data release protocol. Copies of the data are kept and archived, with the original observer generally having exclusive access to the data for the first year or two. After that, the data are generally released to public access, so that any researcher can access the data and use it for their own work.

The Virtual Observatory

Given the ubiquity of data archives at local and national levels, the next step was to make them generally interoperable, and this happens under the general direction of the International Virtual Observatory Alliance (IVOA).

A *virtual observatory* (VO) is the embodiment of an environment that seamlessly accesses data from archives, active missions and other institutional resources worldwide, allowing astronomical research to be conducted. With the capabilities and efficiency of new ground and space-based observatories, this has led to enormous amounts of data being produced. Innovative ways are needed to process and exploit these data (Hanisch et al. 2015; Cui & Zhao 2008; Schaaff 2007).

Consisting of a number of centers having unique collections of astronomical data, software and processing capabilities, a VO can supply researchers access to data on remote computer networks. This is analogous to physical observatories that consist of telescopes and collections of unique astronomical instruments. The difference is the astronomer can have instant access to data using their own computer anywhere. For example, a research team can work with multiwavelength data from the NASA/European Space Agency Hubble Space Telescope, ESO's Very Large Telescope and NASA's Chandra X-ray Observatory in order to do research or write proposals for further data acquisition.

A virtual observatory is comprised of various levels:

- It provides infrastructure (hardware, software and connectivity), coordinates, secures and aggregates necessary project resources.
- It develops specialized resources and tools for seamless data access.
- It offers software/inquiry tools for specific science projects and may communicate with other observatories.

The International Virtual Observatory Alliance (IVOA) was formed in 2002[30] (Peebles 2004), and grew to have 21 member national projects in the subsequent decade. IVOA defines only a general framework and standard for members; each VO maintains their own services and tools depending on their specific goals. In short, IVOA seeks to share astronomical knowledge that is standardized in order to better exploit high-quality data produced around the world.

Virtual Observatory Precursors: Many large-scale data aggregation services predate the virtual observatory, and have been major contributors to astronomical research for decades.

The SIMBAD astronomical database is the world reference database for the identification of astronomical objects and provides basic data, cross-identifications, bibliography, and measurements for astronomical objects outside the solar system. It is integrated with the VizieR catalogue service and the Aladin interactive software

---

[30] This occurred at a conference held in Garching (Germany) named, "Toward an International Virtual Observatory."

sky atlas for access, visualization and analysis of astronomical images, surveys, catalogs, databases and related data, hosted at the CDS (Centre de Données astronomiques de Strasbourg).

The NASA/IPAC Extragalactic Database (NED) is an online astronomical database for astronomers that collates and cross-correlates astronomical information on extragalactic objects (galaxies, quasars, radio, X-ray, and infrared sources, etc), created in the late 1980s. NED is built around a master list of extragalactic objects for which cross-identifications of names have been established, accurate positions and redshifts entered to the extent possible, and some basic data collected. Bibliographic references relevant to individual objects have been compiled, and abstracts of extragalactic interest are kept online.

A Multimessenger Virtual Observatory: The Astrophysical Multimessenger Observatory Network (AMON) is a multimessenger virtual observatory primarily designed to receive and integrate events across all cosmic messengers, including signals that would not usually be strong enough to be detections (sub-threshold events) (Ayala Solares et al. 2020—https://www.amon.psu.edu/).

A Look Ahead

The development of computer hardware, software and networking has driven a model in which it is comparatively easy to have full access to leading-edge astronomical data. New telescope technology, however, is outpacing the capacity of easily-available computer technology. The Australian Square Kilometre Array Pathfinder (ASKAP) radio telescope already requires a supercomputer center for data processing, and the planned Vera Rubin Observatory and Square Kilometre Array (SKA) optical and radio telescopes will have similar computing requirements. It seems likely, therefore, that access to data will become less frictionless over the next decade or two.

References

Ayala Solares, H. A., Coutu, S., Cowen, D. F., et al. 2020, APh, 114, 68

Birney, D. S., Gonzalez, G., & Oesper, D. 2006, Observational Astronomy (Cambridge: Cambridge University Press)

Capova, K. A., Persson, E., Milligan, T., & Dunér, D. 2018, Astrobiology and Society in Europe Today (Berlin: Springer)

Cui, C. Z., & Zhao, Y. H. 2008, in IAU Symp. 248, A Giant Step: from Milli- to Micro-arcsecond Astrometry, ed. W. J. Jin, I. Platais, & M. A. C. Perryman (Cambridge: Cambridge Univ. Press), 563

Dennis, D. W. 2002, AHES, 56, 427

Ford, D. 2014, The Observer's Guide to Planetary Motion (New York: Springer)

Hanisch, R. J., Berriman, G. B., Lazio, T. J. W., et al. 2015, A&C, 11, 190

North, J. 2008, Cosmos: An Illustrated History of Astronomy and Cosmology (Chicago, IL: Univ. Chicago Press)

Peebles, P. J. E. 2004, Toward an International Virtual, ed. P. J. Quinn, & K. M. Górski (Berlin: Springer)

Rogers, L. 2008, It's Only Rocket Science: An Introduction in Plain English (Berlin: Springer)

Schaaff, A. 2007, in Web Information Systems Engineering—WISE 2007 Workshops, ed. M. Weske, M. Hacid, & C. Godart (Berlin: Springer)
Soffel, M., & Langhans, R. 2012, Space-time Reference Systems (Berlin: Springer)
Wells, D. C., Greisen, E. W., & Harten, R. H. 1981, A&AS, 44, 363
Woolard, E. W. 1942, PASP, 54, 77

Principles of Multimessenger Astronomy

Miroslav D Filipović and Nicholas F H Tothill

Appendix A

Metric Prefixes

The International System of Units (SI, abbreviated from the French Systéme International (d'unités)) prefixes used to form decimal multiples and submultiples of SI units are given below in Table A.1 (https://physics.nist.gov/cuu/Units/prefixes. html). The "Treaty of the Metre (Convention du Métre)" was signed in Paris on 1875 May 20 by 48 nations.

The kilogram is the only SI unit with a prefix as part of its name and symbol. With this exception, any SI prefix may be used with any SI unit, including the degree celsius and its symbol °C.

The last letter of a prefix is generally omitted if the first letter of the unit name is a vowel. Thus 100 ares is a hectare and 1 million ohms is a megohm. However, the last letter of the prefix is not omitted if pronunciation is not a problem, as in the case of the milliampere. The letter "l" is sometimes added to prefixes before the erg, so one million ergs is a megalerg.

As the SI prefixes strictly represent powers of 10, they should not be used to represent powers of 2. Thus, one kilobit, or 1 kbit, is 1000 bit and not 2^{10} bit = 1024 bit. To alleviate this ambiguity, separate prefixes for binary multiples have been adopted by the International Electrotechnical Commission (IEC; https://www.iec. ch) for use in information technology.

Table A.1. Metric (SI) Prefixes

Name	Symbol	Value	Name	Symbol	Value
yotta	Y	10^{24}	deci	d	10^{-1}
zetta	Z	10^{21}	centi	c	10^{-2}
exa	E	10^{18}	milli	m	10^{-3}
peta	P	10^{15}	micro	μ	10^{-6}
tera	T	10^{12}	nano	n	10^{-9}
giga	G	10^{9}	pico	p	10^{-12}
mega	M	10^{6}	femto	f	10^{-15}
kilo	k	10^{3}	atto	a	10^{-18}
hecto	h	10^{2}	zepto	z	10^{-21}
deka	da	10	yocto	y	10^{-24}

Principles of Multimessenger Astronomy

Miroslav D Filipović and Nicholas F H Tothill

Appendix B

The Standard SI System of Measurements

The SI is founded on seven SI base units for seven base quantities assumed to be mutually independent, as given in Table B.1. The *cgs*-system is defined in terms of SI units.

Initial definitions:

- The *electro-motive force* (or EMF) of a charge distribution is the work done (in joules) to bring a unit electric charge from infinity up to the distribution.
- The *magneto-motive force* (or MMF) is any physical cause that produces magnetic flux.
- The symbol "AT" stands for Ampere-Turns.

B.1 Definition of the Base Units

The SI base units were redefined in 2019 as follows:

- Ampere: The ampere, symbol A, is the SI unit of electric current. It is defined by taking the fixed numerical value of the elementary charge e to be $1.602176634 \times 10^{-19}$ when expressed in the unit C, which is equal to A s, where the second is defined in terms of $\Delta\nu(Cs)$.
- Meter: The meter, symbol m, is the SI unit of length. It is defined by taking the fixed numerical value of the speed of light in vacuum c to be 299,792,458 when expressed in the unit $m\,s^{-1}$, where the second is defined in terms of the cesium frequency $\Delta\nu(Cs)$.
- Candela: The candela, symbol cd, is the SI unit of luminous intensity in a given direction. It is defined by taking the fixed numerical value of the luminous efficacy of monochromatic radiation of frequency 540×10^{12} Hz, K_{cd}, to be 683 when expressed in the unit $lm\,W^{-1}$, which is equal to $cd\,sr\,W^{-1}$, or $cd\,sr\,kg^{-1}\,m^{-2}\,s^{3}$, where the kilogram, meter and second are defined in terms of h, c and $\Delta\nu(Cs)$.
- Kilogram: The kilogram, symbol kg, is the SI unit of mass. It is defined by taking the fixed numerical value of the Planck constant h to be

Table B.1. The SI of measurement

Quantity	Unit	Symbol	Equivalence
	Base units		
Amount of substance	mole	mol	——
Electric current	ampere	A	——
Length	meter	m	——
Luminous intensity	candela	cd	——
Mass	kilogram	kg	——
Time	second	s	——
Thermodynamic temperature	kelvin	K	——
	Derived units		
Capacitance	farad	F	$s^4 A^2 m^{-1} kg$
Conductance	siemens	S	$s^3 A^2 m^{-2} kg^{-1}$
Electric charge	coulomb	C	$A s$
Electric potential, EMF	volt	V	$m^2 kg s^{-3} A$
Energy, work	joule	J	$m^2 kg s^{-2}$
Frequency	hertz	Hz	s^{-1}
Force	newton	H	$m kg s^{-2}$
Illuminance	lux	lx	$cd sr^a$
Inductance	henry	H	$m^2 kg s^{-2} A^{-2}$
Luminous flux	lumen	lm	$cd sr^a$
Magnetic flux	weber	Wb	$m^2 kg s^{-2} A^{-1}$
Magnetic flux density	tesla	T	$kg s^{-2} A^{-1}$
Pressure	pascal	Pa	$kg m^{-1} s^{-2}$
Power, radiant flux	watt	W	$m^2 kg s^{-3}$
Resistance	ohm	Ω	$m^2 kg s^{-3} A^{-2}$

[a] **Note:** The steradian unit is included to differentiate this unit from the candela.

$6.62607015 \times 10^{-34}$ when expressed in the unit J s, which is equal to $kg\, m^2\, s^{-1}$, where the meter and the second are defined in terms of c and $\Delta\nu(Cs)$.

- The second, symbol s, is the SI unit of time. It is defined by taking the fixed numerical value of the cesium frequency $\Delta\nu(Cs)$, the unperturbed ground-state hyperfine transition frequency of the cesium-133 atom, to be 9,192,631,770 when expressed in the unit Hz, which is equal to s^{-1}.
- Kelvin: The kelvin, symbol K, is the SI unit of thermodynamic temperature. It is defined by taking the fixed numerical value of the Boltzmann constant k to be 1.380649×10^{-23} when expressed in the unit $J\, K^{-1}$, which is equal to $kg\, m^2\, s^{-2}\, K^{-1}$, where the kilogram, meter and second are defined in terms of h, c and $\Delta\nu(Cs)$.
- Mole: The mole, symbol mol, is the SI unit of amount of substance. One mole contains exactly $6.02214076 \times 10^{23}$ elementary entities. This number is the

fixed numerical value of the Avogadro constant, N_A, when expressed in the unit mol^{-1} and is called the Avogadro number. The amount of substance, symbol n, of a system is a measure of the number of specified elementary entities. An elementary entity may be an atom, a molecule, an ion, an electron, any other particle or specified group of particles.

Principles of Multimessenger Astronomy

Miroslav D Filipović and Nicholas F H Tothill

Appendix C

The CGS Measurement System

The SI system of units is based on the older *mks* (meters–kilogram–second) standard of engineering, and has become the standard for most technical publications around the world. This system was designed such that the equations for electromagnetics would be simplified. That is, those equations involving spheres contain 4π, those concerning coils contain 2π and those concerning linear currents do not contain π at all.

Because astronomy involves many calculations with spherical symmetry, it can be argued that the old *cgs* (centimeters–grams–seconds) system of units[1] is still useful. The *cgs* system is still used in major astronomical and astrophysical journals. Thus, we occasionally use the *cgs* system next to SI in this book (Tables C.1 and C.2).

Table C.1. The *cgs*-system of measurement and its relation to the more prevalent SI system. In the *cgs*-system, $\varepsilon_0 = 1$ and $\mu_0 = 1$.

CGS Unit	Sym	Short Description	SI equivalent
abampere	abA	One abampere of current flowing in a circular path of one centimeter radius produces a magnetic field of 2π oersteds at the center of the circle.	$= 10$ A
abcoulomb	abC	One abcoulomb is the charge flowing in one second with a given current of one abampere.	$= 10$ C
abhenry	abH	With one abhenry of inductance, a current changing by one abampere per second (1 abA s^{-1}) produces an electromotive force of one abvolt (1 abV).	$= 10^{-9}$ H

(Continued)

---

[1] Originally proposed by Carl Friedrich Gauss in 1832 and extended by James Clerk Maxwell and William Thomson in 1874.

Table C.1. (*Continued*)

CGS Unit	Sym	Short Description	SI equivalent
abohm	abΩ	A current of one abampere flowing through a resistance of one abohm produces a potential difference of one abvolt (Ohms law).	$= 10^{-9}\Omega$
abvolt	abV	The unit of electromotive force or electric potential difference. When one abvolt exists between two points, one erg of energy is needed to move one abcoulomb of charge between those two points.	$= 10^{-8}$ V
barye	Ba	A unit of pressure equal to one dyne per square centimeter.	$= 0.1$ Pa
centimeter	cm	The basic unit of length in the *cgs*– system.	$= 0.1$ m
dyne	dy	The unit of force (gm cm s^{-2}).	$= 10^{-5}$ N
electrostatic units	esu	see statcoulomb.	—
erg	erg	The unit of energy (gm cm s^{-2}).	$= 10^{-7}$ J
galileo	Gal	Used mainly in gravimetry, equal to one unit of acceleration, 1 Gal = 1 cm s^{-2}.	—
Gauss	G	The unit of magnetic flux density (**B**). One Gauss is defined as one maxwell per square centimeter.	$= 10^{-4}$ T
gilbert	Gi	The unit of magnetomotive force (mmf).	$= \frac{10}{4\pi} AT$
gram	g	The basic unit of mass.	$= 10^{-3}$ kg
maxwell	Mx	The unit of magnetic flux.	$= 10^{-8}$ Wb
oersted	Oe	The unit of magnetic field strength (H). In a vacuum one oersted produces a magnetic flux density of one Gauss. It is also equal to one gilbert per centimeter of flux path.	$= \frac{1000}{\pi}$ AT m^{-1}
poise	P	The unit of dynamic viscosity. Named after Jean Louis Marie Poiseuille. The equivalent SI units are Pascal-seconds.	$= 0.1$ Pa s
second	s	The basic unit of time.	$= 1$ s
statcoulomb	sC	A unit of electric charge. If two particles each carry a charge of one statcoulomb and are one centimeter apart, they repel each other with a force of one dyne.	$= 3.33564 \times 10^{-10}$ C
statOhm	sΩ	See abOhm.	
statvolt	sV	A unit of electric potential. One statvolt per centimeter is equal in magnitude of one Gauss. An electric field of one (statvolt cm^{-1}) has the same energy density as a magnetic field of Gauss.	$= 299.792458$ V
stilb	sb	The unit of luminance.	$= 1$ cd m^{-2}
stokes	St	The unit of kinetic viscosity cm^2 s^{-1}.	$= \frac{1}{10,000}$ m^2 s^{-1}

Table C.2. Some physical constant based in *cgs* and/or SI.

Quantity	Value
Atomic mass unit (amu)	$m_u = 1.6605402 \times 10^{-27}$ kg
Avogadro number	$N_A = 6.0221367 \times 10^{23}$ mol^{-1}
Bohr magneton	$\mu_B = 9.274009994 \times 10^{-24}$ J T^{-1}
Bohr radius	$a_0 = 5.29177210903 \times 10^{-11}$ m
Boltzmann constant	$k = 1.38064852 \times 10^{-23}$ m^2 kg s^{-2} K^{-1}
Classical electron radius	$r_e = 2.8179403227 \times 10^{-15}$ m
Compton electron wavelength	$\lambda_{C,e} = 2.42631023867 \times 10^{-10}$ m
Compton neutron wavelength	$\lambda_{C,n} = 1.319590906\ 8 \times 10^{-15}$ m
Compton proton wavelength	$\lambda_{C_p} = 1.32140985539 \times 10^{-15}$ m
Electron charge	$e = 1.602176634 \times 10^{-19}$ C
Electron volt	eV $= 1.602176634 \times 10^{-19}$ J
Fine structure constant	$1/\alpha = 137.035999\ 084$
Faraday constant	$F = 96,485.332\ 1233100184$ C mol^{-1}
Gravitational constant	$G = 6.67430 \times 10^{-11}$ m^3 kg^{-1} s^{-2}
Magnetic flux quantum	$\Phi_0 = 2.067833848 \times 10^{-15}$ Wb
Mass of electron	$m_e = 9.1093837015 \times 10^{-31}$ kg $m_e c^2 = 0.51099895$ MeV
Mass of neutron	$m_n = 1.67492749804 \times 10^{-27}$ kg $m_n c^2 = 939.56542052$ MeV
Mass of proton	$m_p = 1.67262192369 \times 10^{-27}$ kg $m_p c^2 = 938.27208816$ MeV
Permeability of vacuum	$\mu_0 = 1$ dyne abA^{-2} $\mu_0 = 1.25663706212 \times 10^{-6}$ N A^{-2} $\mu_0 = 4\pi \times 10^{-7}$ N A^{-2} (SI)
Permittivity of vacuum	$\varepsilon_0 = 1$ cm^{-1} $\varepsilon_0 = 8.8541878128 \times 10^{-12}$ F m^{-1} (SI)
Plancks constant	$h = 6.6260755 \times 10^{-27}$ erg s $h = 4.135667696 \times 10^{-15}$ eV s $h = 6.62607015 \times 10^{-34}$ J s (SI)
Rydberg constant	$R_\infty = 10,973,731.568160$ m^{-1} $R_\infty c = 3.2898419602508 \times 10^{15}$ Hz $h\,c\,R_\infty = 13.605693122994$ eV
Speed of light in vacuum	$c = 299792458$ m s^{-1} (SI)
Stefan–Boltzmann constant	$\sigma = 5.67051 \times 10^{-5}$ erg cm^{-2} K^{-4} s^{-1} $\sigma = 5.670374419$ W m^{-2} K^{-4} (SI)
Thomson cross section	$\sigma_e = 6.6524587321 \times 10^{-29}$ m^2

Appendix D

Unit Conversions

X-ray emissions are sometimes measured in milliCrabs (mCr)[†]. As the name implies, this unit uses the X-ray emissions from the (relatively) near-by Supernova Remnant Crab Nebula (a.k.a. M 1 and SN 1054) as a standard. This unit is defined in the following way:

$$I(\mathrm{mCr}) = 10^3 \frac{\int_{E1}^{E2} E\left(\frac{dN}{dE}\right) dE}{\int_{E1}^{E2} E\left(\frac{dN}{dE}\right)_{\mathrm{Crab}} dE} \tag{D.1}$$

where $\frac{dN}{dE}$ and $\left(\frac{dN}{dE}\right)_{\mathrm{Crab}}$ are the source and Crab Nebula photon spectral flux density, respectively. Using:

$$\left(\frac{dN}{dE}\right)_{\mathrm{Crab}} = 10E(\mathrm{keV})^{-2.05} \exp\left[-\sigma N_H\right] \text{photons } \mathrm{cm}^{-2}\,\mathrm{s}^{-1}\,\mathrm{keV}^{-1} \tag{D.2}$$

with $N_H = 3 \times 10^{21}$ cm^{-2} and σ is the absorption and scattering cross-section. Then, setting $E1 = 2$ keV and $E2 = 10$ keV, we get the 2.4×10^{-8} erg cm^{-2} s^{-1}.

The X-ray luminosity in the range of 0.1–100 keV is $L = 4.9 \times 10^{37}$ erg s^{-1}. See also, Handbook of Space Astronomy and Astrophysics by Martin Zombeck, from Cambridge Univ. Press.

Table D.1. Some Common Unit Conversions

From = To	Comments
Distance or length	
1 pc = 3.261633 lt-yr	
1 pc = 3.085678×10^{16} m	
1 light-year = 9.460530×10^{15} m	
1 Å = 10^{-10} m	
Energy	
1 eV = $1.602176\,634 \times 10^{-16}$ J	
1 eV = 1.97327×10^{-7} m	(hc/E)
1 eV = 1.160451812×10^4 K	(E/k)
1 erg = 10^{-7} J	
Pressure	
760 torr = 1.013×10^6 dyne cm^{-2} = 1 atmosphere	
760 torr = 1.01325 bars = 101,325 pascals	
1 pascal = 10^{-5} bars = 0.00131579 atmosphere	
Energy flux	
1 Rayleigh = 1/4 $\pi \times 10^6$ photons cm^{-2} s^{-1} sr^{-1}	
1 Uhuru s^{-1} = 1.7×10^{-11} erg cm^{-2} s^{-1}	Range: (2–6 keV)
1 Uhuru s^{-1} = 2.4×10^{-11} erg cm^{-2} s^{-1}	Range: (2–10 keV)
1 jansky (Jy) = 10^{-23} erg s^{-1} cm^{-2} Hz^{-1}	Also equal to one flux unit.
1 jansky (Jy) = 10^{-26} W m^{-2} Hz^{-1}	
1 mCr = 2.4×10^{-8} erg cm^{-2} s^{-1}	MilliCrab units of X-ray
1 mCr = 15 keV cm^{-2} s^{-1} = 2.4×10^{-11} W m^{-2}	emission [†]. Range: (2–10 keV)

Principles of Multimessenger Astronomy

Miroslav D Filipović and Nicholas F H Tothill

Appendix E

Distributions

This appendix describes a few statistical distributions relevant to spectroscopy and astronomy in general.

E.1 The Airy Function

The Airy function describes the distribution of light in the image of a star formed by a telescope. The general form of the Airy function is:

$$Ai(x) = \frac{1}{\pi} \int_0^\infty \cos\left[\frac{t^3}{3} + xt\right] dt. \tag{E.1}$$

E.2 Lorentzian Distribution

The Lorentzian distribution describes the shape of a spectral line as it is broadened by a number of mechanisms, in particular natural or collisional broadening. In mathematics this function is known as the Cauchy[1] distribution. The function is:

$$f(\nu, \nu_0, \omega) = \frac{1}{\pi}\left[\frac{\omega}{(\nu - \nu_0)^2 + \omega^2}\right] \tag{E.2}$$

where μ_0 is the center frequency of the spectral line and ω is the half-width at half-maximum of the line.

E.3 Lévy Skew Alpha-stable Distribution

The Lévy skew alpha-stable distribution[2] is actually a family of distributions, one of which is the Van der Waals distribution, which describes the spectral line shape for broadening by Van der Waals forces. The Gaussian and Cauchy distributions can be derived from the Lévy skew alpha-stable distribution.

---

[1] Augustin Cauchy (1789–1857).
[2] Paul Pierre Lévy (1886–1971).

Appendix F

The Electromagnetic Spectrum

The electromagnetic spectrum has been divided into regions and bands. The regions, from high frequency to low are:

- Gamma-ray, or γ-ray (> 3 EHz),
- X-ray (30 EHz–30 PHz),
- Ultraviolet (30 PHz–750 THz),
- Optical (750 THz–430 THz),
- Infrared (430 THz–3 THz),
- Submillimeter (3 THz–300 GHz),
- Microwave (300 GHz–3 GHz) and
- Radio (< 3 GHz).

F.1 Regions and Bands of the Electromagnetic Spectrum

The regions of the spectrum are further divided into bands. Tables F.1–F.10 show the spectral regions and their bands as used in astronomy. In the higher frequency bands (UV, X-ray, and gamma-ray), the borders separating the sub-bands are not well defined. For example, EUV (extreme ultraviolet) has some overlap with soft X-rays. The main criteria used to distinguish soft and hard X-rays, and soft and hard gamma-rays, and indeed, to distinguish X-rays from gamma-rays, is the method of detection and the source of the radiation. These tables use some less than well-known metric prefixes (Appendix A).

A region between ultraviolet and X-rays (25 Å–912 Å 120 PHz–3.3 PHz, 496 eV–13.6 eV) is highly absorbed by the interstellar medium and is practically unobservable (see Figure 5.16).

Near UV is often included as part of the optical band, but it is hindered by poor reflectivity or transmission of optical elements, and atmospheric opacity. The border for vacuum UV is established by the opacity of air to ultraviolet. Extreme UV is characterized by a physical transition between photons interacting mainly with the valence electrons, and photons interacting mainly with the core electrons and the nucleus of atoms.

doi:10.1088/2514-3433/ac087ech15

Table F.1. The γ-ray Region

ν (Hz)	λ (m)	E (eV)	Desig.	Comments
< 24 E	< 12 p	> 100 k		All γ-rays
> 7.3 Z	<41 f	> 30 M	Hard γ-rays	Pair production detectors
7.3 Z–24 E	41 f–12 p	30 M–100 k	Soft γ-rays	Scintillation detectors

Table F.2. The X-ray Region

ν (Hz)	λ (m)	E (eV)	Desig.	Comments
24 E–30 P	12 p–12 n	100–0.1 k		All X-rays
24 E–2.4 E	12 p–0.12 n	100–10 k	Hard X-Ray	Scintillation detectors
2.4 E–30 P	0.12 n–12 n	10–0.1 k	Soft X-ray	Grazing incidence mirrors

Table F.3. The Ultraviolet Region

ν (Hz)	λ (nm)	E (eV)	Desig.	Comments
300 P–749 T	1–400	1240–3.10		All UV
300 P–30 P	1–10	1240–124	EUV, XUV	Extreme UV
30 P–3 P	10–100	124–12.4	FUV, VUV	Far or vacuum UV
3 P–1.07 P	100–280	12.4–4.43	UVC	Near UV, "germicidal"
1070 T–952 T	280–315	4.43–3.94	UVB	Near UV
952 T–749 T	315–400	3.94–3.10	UVA	Near UV, "black light"

Because infrared radiation is associated with heat energy, this spectral region is described both in terms of wavelength and in terms of the equivalent blackbody temperature T (Section 2.2.6), for that wavelength. The conversion is based on Wiens law:

$$T = \frac{2.89 \times 10^{-3} \text{ mK}}{\lambda} = \frac{2898 \text{ μm K}}{\lambda}. \tag{F.1}$$

Infrared wavelengths are mostly in micrometers (μm) or "microns." This is often abbreviated to just μ. Short wavelength, near-infrared can be observed with optical telescopes (using appropriate, cooled detectors). Longer wavelengths require more specialized telescope constructions. The following infrared bands are passed through the atmosphere and thus allow ground based infrared astronomy (on high mountain tops): 1.1–1.4 μm, 1.5–1.8 μm, 2.0–2.4 μm, 3.0–4.0 μm, 4.6–5.0 μm, 7.5–14.5 μm,

Table F.4. The Optical Region: Spectral Lines of Interest for Astronomical Observation

ν (THz)	λ (nm)	T (K)	Desig.	Comments
749–411	400–730	7230–3960		All optical
730.8	410.2	7045	Balmer Series	H_δ
690.6	434.1	6660		H_γ
616.7	486.1	5950		H_β
4576.8	656.3	4400		H_α
UBV (RI) filter system				
817	367	7870	Ultraviolet (U)	BW = 66 nm
688	436	6630	Blue (B)	BW = 94 nm
550	545	5300	Visible (V)	BW = 88 nm
470	638	4530	Red (R)	BW = 138 nm
376	797	3630	Infrared (I)	BW = 149 nm
Uvby filter system				
857	350	8260	Ultraviolet	BW = 30 nm
730	411	7030	Visible	BW = 19 nm
642	467	6190	Blue	BW = 18 nm
548	547	5280	Yellow	BW = 23 nm
Johnson–Cousins–Glass filter system				
246	1220	2370	Infrared (J)	BW = 213 nm
184	1630	1770	Infrared (H)	BW = 307 nm
137	2190	1320	Infrared (K)	BW = 390 nm
87	3450	838	Infrared (L)	BW = 472 nm
63	4750	608	Infrared (M)	BW = 460 nm

Note. 1 nm = 10 Å.

Table F.5. The Infrared (IR) Region

ν (THz)	λ (μm)	T (K)	Desig.	Comments
411–0.86	0.73–350			All infrared
411–60	0.73–5.0	3950–578	Near IR	Cooler red stars
				Red giants
				Dust is transparent
60–10	5.0–30	578–96	Mid IR	Planet, comets, asteroids
				Dust warmed by starlight
				Protoplanetary disks
10–0.86	30–350	96–8.3	Far IR	Emissions from cold dust
				Central regions of galaxies
				Cold molecular clouds

Table F.6. The Submillimeter Bands

ν (THz)	λ (mm)	Desig.	Comments
3.0–0.3	0.1–1	Submillimeter	Detected by radio telescopes

Table F.7. The General Microwave Bands

ν (GHz)	λ (cm)	Desig.	Comments
300–3	0.1–10		All microwaves
300–30	0.1–1	EHF	Extra high frequency
30–3	1–10	SHF	Super high frequency
110–75	0.27–0.4	W	
75–0	0.4–0.75	V	
40–26	0.75–1.2	kA	
26–18	1.2–1.7	K	
18–12	1.7–2.5	Ku	
12–8	2.5–3.75	X	
8–4	3.75–7.5	C or Q	
4–2	7.5–15	S	μ wave/Radio
2–0.5	15–60	L	μ wave/Radio
0.5–0.25	60–120	P	μ wave/Radio
0.25–0.2	120–150	G	μ wave/Radio
< 0.2	> 150	I	μ wave/Radio

Table F.8. The Microwave and Radio Bands Allocated to Astronomy

ν (GHz)	λ (mm)	Desig.	Comments
95.00–100.0	3.156–2.998	W-Band	
92.00–94.00	3.259–3.189		
86.00–92.00	3.486–3.259		
72.77–72.91	4.120–4.112	V-Band	Formaldehyde line
58.20–59.00	5.151–5.081		
51.40–54.25	5.833–5.526		
48.94–49.04	6.126–6.113	C-Band	Carbon monosulphide
42.50–43.50	7.054–6.892		Silicon monoxide
36.43–36.50	8.229–8.214	Ka-Band	Hydrogen–cyanide & Hydroxyl
31.20–31.80	9.609–9.427		
23.60–24.00	12.70–12.49		Ammonia line
23.07–23.12	12.99–12.97		
22.86–22.91	13.11–13.09		Ammonia
22.01–22.50	13.62–13.32		Water lines
15.35–15.40	19.53–19.47	Ku-Band	
15.20–15.35	19.72–19.53		

14.47–14.50	20.72–20.68	
10.06–10.70	28.28–28.02	X-Band
6.6500–6.6752	45.082–44.911	C-Band
5.000 0–5.0300	59.959–59.601	
4.990 0–5.0000	60.079–59.959	
4.8000–4.9900	62.457–70.079	
3.345 8–3.3525	89.603–89.424	S-Band
3.3320–3.3390	89.974–89.785	
3.2600–3.2670	91.961–91.764	
2.6900–2.7000	111.45–111.03	
2.6550–2.6900	112.92–111.45	
1.7188–1.7222	174.42–174.08	L-Band
1.6684–1.6700	179.69–179.52	
1.6605–1.6684	180.54–179.69	
1.6600–1.6605	180.60–180.54	
1.4000–1.4270	214.10–210.10	

Table F.9. These Radio Bands are Protected for Astronomical Observation

ν (Hz)	λ (m)	Desig.	Comments
300 M–3 G	1–10 c	UHF	Ultra high frequency
30 M–300 M	10–1	VHF	Very high frequency
3 M–30 M	100–10	HF	High frequency
300 k–3 M	1 k–100	MF	Medium frequency
30 k–300 k	10 k–1 k	LF	Low frequency
3 k–30 k	100 k–10 k	VLF	Very low frequency
300–3 k	1000 k–100 k	VF	Voice frequency
30–300	10,000 k–1000 k	ELF	Extra low frequency
< 30	> 10,000 k	ULF	Ultra low frequency

Table F.10. These are Recognized Bands for General Radio Transmission

ν (MHz)	λ (m)	Desig.	Comments
608.0–614.0	0.493–0.488		
406.0–410.0	0.738–0.731		Pulsar observations
322.0–328.6	0.931–0.912	UHF	
150.05–153.00	1.998–1.959		Pulsar observations Solar observations
73.00–74.60	4.11–4.02		Solar wind observations
37.50–38.25	7.99–7.84	VHF	
25.55–25.67	11.73–11.68		
13.36–13.41	22.44–22.35	HF	

17–40 µm, 330–370 µm. These narrow bands have varying transparency and trouble with sky brightness, with longer wavelengths having worse performance.

The designation of wavelength bands in the microwave region has a long and confused history. Here is at least one version we assembled from a variety of sources. At the lower end of these bands, there is an overlap with the radio bands. This is probably due to historical issues.

Appendix G

Examples of Astronomy Software and Tools

TOPCAT (http://www.star.bris.ac.uk/~mbt/topcat/): TOPCAT is an interactive graphical viewer and editor for tabular data. Its aim is to provide most of the facilities that astronomers need for analysis and manipulation of source catalogs and other tables, though it can be used for non-astronomical data as well. It is especially good at interactive exploration of large (multimillion row, lots of columns) tables. It offers a variety of ways to view and analyze tables, including a browser for the cell data themselves, viewers for information about table and column metadata, and facilities for sophisticated interactive and higher-dimensional visualization, calculating statistics, and joining tables using flexible matching algorithms.

It is a stand-alone application which works quite happily without a network connection. However, because it uses Virtual Observatory (VO) standards, it can cooperate smoothly with other tools, services and data sets in the VO world and beyond. It has been developed mostly in the UK within various UK and Euro-VO projects (Starlink, AstroGrid, VOTech, AIDA, GAVO, GENIUS, DPAC).

Aladin (https://aladin.u-strasbg.fr/): Created in 1999, Aladin has become a widely used VO portal capable of addressing challenges such as locating data of interest, accessing and exploring distributed data sets, visualizing multiwavelength data. Compliance with existing or emerging VO standards, interconnection with other visualization or analysis tools, ability to easily compare heterogeneous data are key topics allowing Aladin to be a powerful data exploration and integration tool as well as a science enabler.

SIMBAD (http://simbad.u-strasbg.fr/): SIMBAD astronomical database is the world reference database for the identification of astronomical objects and provides basic data, cross-identifications, bibliography and measurements for astronomical objects outside the solar system. Using VizieR, it is the catalog service for the CDS (Centre de Données astronomiques de Strasbourg) reference collection of astronomical catalogs and tables published in academic journals and the Aladin interactive software sky atlas for access, visualization and analysis of astronomical images, surveys, catalogs, databases, and related data. Simbad bibliographic survey began in 1950 for stars (bright stars) and in 1983 for all other objects (outside the solar system).

VizieR (https://vizier.u-strasbg.fr/): The CDS had for years collected and disseminated astronomical data, so the original ESIS Catalog Browser was transferred to and stored there. Since its inception in 1996, VizieR has become a reference point for astronomers worldwide engaged in research, who come to access cataloged data regularly published in astronomical journals.

NED (http://ned.ipac.caltech.edu/): The NASA/IPAC Extragalactic Database (NED) is an online astronomical database for astronomers that collates and cross-correlates astronomical information on extragalactic objects (galaxies, quasars, radio, X-ray, and infrared sources, etc). NED was created in the late 1980s and is funded by NASA. NED is built around a master list of extragalactic objects for which cross-identifications of names have been established, accurate positions and redshifts entered to the extent possible, and some basic data collected. Bibliographic references relevant to individual objects have been compiled, and abstracts of extragalactic interest are kept online.

DS9 (http://SAOImageDS9ds9.si.edu): SAOImage DS9 is an astronomical imaging and data visualization application. DS9 supports Flexible Image Transport System (FITS) images and binary tables, multiple frame buffers, region manipulation, and many scale algorithms and colormaps. It provides for easy communication with external analysis tasks and is highly configurable and extensible. DS9 supports advanced features such as 2-D, 3-D and RGB frame buffers, mosaic images, tiling, blinking, geometric markers, colormap manipulation, scaling, arbitrary zoom, cropping, rotation, pan, and a variety of coordinate systems.

KARMA (https://www.atnf.csiro.au/computing/software/karma/): Karma is an astronomical software package for visualizing multidimensional images, signal and image processing applications.

G.1 Other Astronomy Community Software Resources

- Astropy—Community-based Python astronomy software project
    https://www.astropy.org/
- AstroConda—STScI astronomy software distribution
    https://astroconda.readthedocs.io/en/latest/
- Gemini Data Reduction Software
    https://www.gemini.edu/observing/phase-iii
- Gemini DRAGONS project (next generation Gemini data reduction software)
    https://github.com/GeminiDRSoftware/DRAGONS
- IRAF—Image Reduction and Analysis Facility
    http://ast.noao.edu/data/software
- CFITSIO—FITS Library and compression tools
    https://heasarc.gsfc.nasa.gov/fitsio/
- WCSTOOLS—Image header utilities
    http://tdc-www.harvard.edu/wcstools/

- STSDAS/tables—IRAF Packages from STScI
- PyRAF—Python-based alternative to the IRAF CL
    https://www.stsci.edu/itt/review/dhb_2011/Intro/intro_ch45.html
- AIPS—Astronomical Image Processing System, produced by NRAO
    http://www.aips.nrao.edu/index.shtml
- MIRIAD—ATNF's radio interferometry data reduction package
    https://www.atnf.csiro.au/computing/software/miriad/
- CASApy—Common Astronomy Software Applications—the full NRAO package
    https://casa.nrao.edu/gettingstarted.shtml
- CARTA—Cube Analysis and Rendering Tool for Astronomy, is a next generation image visualization and analysis tool designed for Atacama Large Milli meter/submillimeter Array (ALMA), VLA and SKA pathfinders
    https://cartavis.github.io/

CPSIA information can be obtained
at www.ICGtesting.com
Printed in the USA
BVHW020531080122
625739BV00004B/50

9 780750 323383